Brains, Machines, and Mathematics

Michael A. Arbib

Brains, Machines, and Mathematics

Second Edition

With 63 Illustrations

Springer-Verlag
New York Berlin Heidelberg
London Paris Tokyo

Michael A. Arbib
Departments of Biomedical Engineering,
    Computer Science, Electrical Engineering,
    Neurobiology, Physiology, and Psychology
University of Southern California
Los Angeles, CA 90089-0782
USA

Library of Congress Cataloging-in-Publication Data
Arbib, Michael A.
    Brains, machines, and mathematics.
    Bibliography: p.
    Includes index.
    1. Cybernetics.   2. Machine theory.   I. Title.
Q310.A7   1987   001.53   87-9806

Previous edition: M.A. Arbib, *Brains, Machines, and Mathematics.*
© 1964 by McGraw-Hill, Inc.

© 1987 by Springer-Verlag New York Inc.

Typeset by Asco Trade Typesetting Ltd., Hong Kong.
Printed and bound by R.R. Donnelley & Sons, Harrisonburg, Virginia.
Printed in the United States of America.

9 8 7 6 5 4 3 2 1

ISBN 0-387-96539-4 Springer-Verlag New York Berlin Heidelberg
ISBN 3-540-96539-4 Springer-Verlag Berlin Heidelberg New York

To
Fred Pollock and Rhys Jones
for teaching wisely, wittily, and well

Preface to the Second Edition

This is a book whose time has come—again. The first edition (published by McGraw-Hill in 1964) was written in 1962, and it celebrated a number of approaches to developing an automata theory that could provide insights into the processing of information in brainlike machines, making it accessible to readers with no more than a college freshman's knowledge of mathematics. The book introduced many readers to aspects of cybernetics—the study of computation and control in animal and machine. But by the mid-1960s, many workers abandoned the integrated study of brains and machines to pursue artificial intelligence (AI) as an end in itself—the programming of computers to exhibit some aspects of human intelligence, but with the emphasis on achieving some benchmark of performance rather than on capturing the mechanisms by which humans were themselves intelligent. Some workers tried to use concepts from AI to model human cognition using computer programs, but were so dominated by the metaphor "the mind is a computer" that many argued that the mind must share with the computers of the 1960s the property of being serial, of executing a series of operations one at a time. As the 1960s became the 1970s, this trend continued. Meanwhile, experimental neuroscience saw an exploration of new data on the anatomy and physiology of neural circuitry, but little of this research placed these circuits in the context of overall behavior, and little was informed by theoretical concepts beyond feedback mechanisms and feature detectors. By contrast, I have always insisted that the metaphor "the brain is a computer" must not be read as reducing the brain to the level of current computer technology, but rather must expand our concepts of computation to embrace the style of the brain, depending on the constant interaction of many concurrently active systems, many of which express their activity in the interplay of spatiotemporal patterns in diverse layers of neurons.

However, many computer scientists are now abandoning the serial computer paradigm, and are seeking to understand how problem-solving can be distributed across a network of interacting concurrently active processors, while within AI a whole new field of "connectionism" or PDP (parallel distributed processing) has arisen, which combines the insight of studies of "learning networks" made in the 1950s and 1960s with the sophistication gained from more than two decades of studies in AI.

Chapter 1 provides a historical overview tracing the rise of cybernetics, its dissolution into a number of specialized disciplines, and the new rapprochement that makes the Second Edition of *Brains, Machines, and Mathematics* so timely. Chapter 2 then presents the classic two-state model of the neuron due to McCulloch and Pitts and shows how any finite-state machine can be simulated by a network of such neurons. Chapter 3 introduces the crucial cybernetic concept of feedback and analyzes a problem crucial to the adaptive control of a system, namely, the realization problem of inferring a state-space description of the system from samples of its behavior. Chapter 4 then looks at the role of layered networks of neurons in pattern recognition, introducing two classic schemes for making such networks "learn from experience": Rosenblatt's Perceptron and Hebb's scheme for unsupervised learning. The chapter closes by presenting results due to Minsky–Papert and Winograd–Spira, which show how limitations on network topology and neuron complexity limit what a network can do, irrespective of whatever learning rule is employed.

It must be stressed that the emphasis here is on automata theory, analyzing the capabilities of networks of simple components similar to McCulloch–Pitts neurons equipped with some "learning rule" to adjust their connections. These are networks "in the style of the brain," rather than models responsible to the current data on cellular neuroanatomy and neurophysiology. This story is continued when we examine a number of studies in the new connectionist or PDP paradigm, which builds AI systems or models cognitive functions in terms of such highly parallel networks. Chapter 5 presents a sampling of current studies of "semi-neural" learning networks, specifically studies of synaptic matrices, Hopfield nets, Boltzmann machines, reinforcement learning, and an interesting back-propagation algorithm. It is in the companion volume *The Metaphorical Brain* that I present realistic models of the regions of real brains, in addition to related studies in machine vision and robotics.

In 1936 (when the word "computer" denoted a *person* carrying out computations) Alan Turing introduced a model of a computer as a formal machine that could be programmed to carry out any effective procedure, i.e., to do anything that a human could do in the way of symbol manipulation when constrained to follow a well-specified set of instructions. Turing's research was part of an effort by mathematical logicians in the 1930s to delimit what could and could not be achieved by formal procedures. The classic 1943 paper of McCulloch and Pitts showed that networks of their formal neurons could provide the "brain" for any Turing machine. Having devoted several chapters to the sequels to McCulloch and Pitts, we devote the last three chapters to research rooted in the logic of the 1930s, charting the limits of the computable whether or not the computation is implemented in a neural network.

Chapter 6 focuses on the capabilities of Turing machines, including the demonstration that there is a *universal* Turing machine which, when suitably programmed, can simulate any other Turing machine. It also demonstrates the unsolvability of the halting problem, i.e., that there is no program to tell whether an arbitrary machine will eventually halt computation. Chapter 7 studies von Neumann's extension of Turing's universal computer to a universal constructor, and then examines a variety of approaches to self-reproducing machines, all in the framework of cellular or tesellation automata. It is ironic that today many computer scientists use the term "von Neumann computer" to refer to serial computer architectures, for von Neumann's work on cellular automata makes him a pioneer in "non-von Neumann architecture"!

Finally, Chapter 8 provides two accessible proofs of Gödel's celebrated Incompleteness Theorem, and deepens the reader's understanding by contrasting it with Gödel's Completeness Theorem, for which a proof is also given. We include a number of little known results, such as that the incompleteness found by Gödel can be removed incrementally in a mechanical fashion, and use this deeper understanding to argue against those philosophers who believe that Gödel's Incompleteness Theorem places inherent limitations, which are not shared by humans, on the possible intelligence of machines.

Having thus outlined the present volume, I now briefly present its relation to five other volumes to be published at much the same time. *The Metaphorical Brain* (Michael A. Arbib, 2nd edition, Wiley-Interscience, 1988) presents my approach to brain theory in which perceptual and motor schemas provide the functional analysis of behavior, while layered neural networks provide models for neural mechanisms of visuomotor coordination that can be put to experimental test. *Neural Models of Visuomotor Coordination* (Michael A. Arbib, Rolando Lara, Francisco Cervantes-Perez, and Donald House, in preparation) will provide a much more detailed look at our research on modeling the mechanisms of visuomotor coordination, especially in frog and toad; on pattern recognition, depth perception, and detour behavior; and reviewing experimental data and offering a careful analysis of both mathematical and computer models. *From Schema Theory in Language* (Michael A. Arbib, E. Jeffrey Conklin, and Jane C. Hill, Oxford University Press, 1987) develops a schema-theoretic approach to language that is exemplified by three studies: of the effects of brain damage on sentence comprehension, of language acquisition by a two-year-old child, and of the use of salience in generating verbal descriptions. In each case, the result is a computational model linked to cognitive studies of human language performance. *The Construction of Reality* (Michael A. Arbib and Mary B. Hesse, Cambridge University Press, 1986) synthesizes my individualist view of schemas as "units of knowledge in the head" with Professor Hesse's view of knowledge as a social construct shared, e.g., by a community of scientists or of religious believers. We thus develop and extend the view of knowledge, including language, as inherently metaphorical and dynamic. Within this shared perspective, we then debate such issues as the freedom of the will and the reality of God. Finally, *In Search of the Person* (Michael A. Arbib, University of Massachusetts Press, 1985) is a small volume, designed to be read in an evening or two, either for itself alone or as an invitation to the weightier development of its themes in *The Construction of Reality*. It is subtitled "Philosophical Explorations in Cognitive Science."

Much of Chapter 5 was written while I was a Visiting Scholar at the Institute for Cognitive Science at the University of California at San Diego (UCSD) during the Academic Year 1985–1986, on leave from the University of Massachusetts at Amherst. I would like to thank Andy Barto for his invaluable survey of the material for me before I left Amherst and for his vital help with the final draft, as well as to thank David Rumelhart, David Zipser, Gary Cottrell, Ron Williams, and other members of the PDP group for their scientific hospitality and insightful discussions during my stay at UCSD, and Geoffrey Hinton and Terry Sejnowski for their helpful comments on the semifinal draft. Finally, I thank Darlene Freedman for her excellent help in bringing the manuscript to publication.

Los Angeles, CA   Michael A. Arbib
August 1986

From the Preface to the First Edition

This book forms an *introduction* to the common ground of brains, machines, and mathematics, where mathematics is used to exploit analogies between the working of brains and the control – computation – communication aspects of machines. It is designed for a reader who has heard of such currently fashionable topics as cybernetics and Gödel's theorem and wants to gain from one source more of an understanding of them than is afforded by popularizations. Here the reader will find not only *what* certain results are, but also *why*. A lot of ground is covered, and the reader who wants to go further should find himself reasonably well prepared to go on to study the technical literature. To make full use of this book requires a moderate mathematical background—a year of college mathematics (or the equivalent "mathematical maturity"). However, much of the book should be intelligible to the reader who chooses to skip the mathematical proofs, and no previous study of biology or computers is required at all.

The use of microelectrodes, electron microscopes, and radioactive tracers has yielded a huge increase in neurophysiological knowledge in the past few decades. Even a multivolume work such as the *Handbook of Neurophysiology* cannot fully cover all the facts. Many neurophysiological theories, once widely held, are being questioned as improved techniques reveal finer structures and more sophisticated chemicoelectrical cellular mechanisms. This means that our presentation of mathematical models in this book will have to be based on a grossly simplified view of the brain and the central nervous system. The reader may well begin to wonder what value or interest the study of such systems can have.

There is a variety of properties—memory, computation, learning, purposiveness, reliability despite component malfunction—which it might seem difficult to attribute to "mere mechanisms." However, herein lies one important reason for our study: By making mathematical models, we have proved that there do exist purely electrochemical mechanisms that have the above properties. In other words, we have helped to "banish the ghost from the machine." We may not *yet* have modeled *the* mechanisms that the brain employs, but we have at least modeled *possible* mechanisms, and that in itself is a great stride forward.

There is another reason for such a study, and it goes much deeper. Many of the most spectacular advances in *physical* science have come from the wedding of the mathematicodeductive method and the experimental method. The mathematics of the last 300 years has grown largely out of the needs of physics—applied mathematics directly, and pure mathematics indirectly by a process of abstraction from applied mathematics (often for purely esthetic reasons far removed from any practical considerations). In these pages we coerce what is essentially still the mathematics of the physicist to help our slowly dawning comprehension of the brain and its electromechanical analogs.

It is probable that the dim beginnings of *biological* mathematics here discernible will one day happily bloom into new and exciting systems of pure mathematics. Here, however, we *apply* mathematics to derive far-reaching conclusions from clearly stated premises. We can test the adequacy of a model by expressing it in mathematical form and using our mathematical tools to prove general theorems.

In the light of any discrepancies we find between these theorems and experiments, we may return to our premises and reformulate them, thus gaining a deeper understanding of the workings of the brain. Furthermore, such theories can guide us in building more useful and sophisticated machines.

The beauty of this mathematicodeductive method is that it allows us to *prove* general properties of our models and thus affords a powerful adjunct to model making in the wire and test-tube sense.

Biological systems are so much more complicated than the usual systems of physics that we cannot expect to achieve a fully satisfactory *biological* mathematics for many years to come. However, the quest is a very real and important one. This book strives to introduce the reader to its early stages. He or she will, I hope, find that the results so far obtained are of interest. Certainly they represent only a very minute fraction of what remains to be found—but the start of a quest is nonetheless exciting for being the start. I do not believe that the application of mathematics will solve all our physiological and psychological problems. What I do believe, though, is that the mathematicodeductive method must take an important place beside the experiments and clinical studies of the neurophysiologist and the psychologist in our drive to understand brains, just as it has already helped the electrical engineer to build the electronic computers that, although many, many degrees of magnitude less sophisticated than biological organisms, still represent our closest man-made analog to brains.

This book is a revision of the lecture notes of a course delivered in June–August of 1962 at the University of New South Wales in Sydney, Australia. I want to thank John Blatt for inviting me to the Visiting Lectureship; Derek Broadbent for inviting me to broadcast the lectures; and Joyce Kean for her superb job of typing up the original lecture notes.

I have spent the last two years with the Research Laboratory of Electronics and the Department of Mathematics at the Massachusetts Institute of Technology on a research assistantship (supported by the U.S. Armed Forces and National Institute of Health). [I was a graduate student at MIT from January of 1961 till September of 1963.] I owe so many debts to the people there that I cannot fully do justice to them. I do particularly want to thank Warren McCulloch for his continual help and encouragement. It was George W. Zopf who first urged publication of the lectures. Bill Kilmer gave the original lecture notes a helpful and critical reading. For years I have nurtured the desire to claim at a point such as this, "Any mistakes which remain are thus solely his responsibility." However, this would be a sorry expression of a very genuine gratitude, and so I follow convention and admit that any errors that remain are my responsibility.

Finally I should like to thank all those authors whose work I have quoted and their publishers for so graciously granting me permission to use their material.

M.A.A.

Contents

CHAPTER 1

A Historical Perspective

Many of the questions addressed in Artificial Intelligence (AI) and Brain Theory in fact have a long history. The aim of this chapter is to present briefly some of that history. Section 1, The Road to 1943, traces the story to 1943, which saw the publication of three remarkable papers: McCulloch and Pitts giving a logical theory of neural networks; Rosenbleuth, Wiener, and Bigelow asserting that a machine with feedback is imbued with purpose; while Craik saw the ability of the brain to simulate the world as providing the key to intelligence. Section 2, Cybernetics Defined and Dissolved, shows how Cybernetics emerged from these studies, only to give birth to a number of distinct new disciplines—such as AI, biological control theory, cognitive psychology, and neural modeling—which each went their separate ways, and shows how brain theory arose therefrom. Section 3 then charts the new rapprochement between AI and Brain Theory that gives new solutions to the cybernetic concerns of the 1940s and 1950s. We first note briefly the rapprochement among AI, cognitive psychology, and linguistics, which brought them together under the banner of cognitive science, and then see how the increasing concern of workers in AI and cognitive psychology with parallelism led to the development of the style known as "connectionism" or PDP (parallel distributed processing), which fosters the rapprochement with brain theory.

1.1 The Road to 1943

In this section, we study the prehistory of cybernetics. But before doing this, we note that the word itself has an interesting history, coming from the Greek word *kybernetes*, meaning the helmsman of a ship. In Latin, the word took on a metaphorical meaning, and the pronunciation "governor." In 1848, Ampère used the word in this political sense when he wrote a treatise on *la cybernetique*. However, as we shall see below, Wiener returned to the original sense when he made feedback (cf. Section 3.1) central to his conception of cybernetics. Feedback is the process, e.g., whereby the helmsman notes the "error," the extent to which he is off course, and "feeds it back" to decide which way to move the rudder. While on the subject of etymology, we may note that the word "automaton" (plural, automata), often used to describe computers and other complex machines, comes from the same Greek root as automobile, meaning "self-mover." A notable early example of an automaton is that designed by Hero of Alexandria around 11 AD as a vending machine for holy water.

Technology has always played a crucial role in attempts to understand the human mind and body; for example, the study of the steam engine has contributed concepts to the study of metabolism, and electricity has been part of the study of the brain at least since Galvani touched frog leg to iron railing sometime before 1791. In 1748 La Mettrie published *L'Homme Machine* and suggested that such automata as the mechanical duck and flute player of Vaucanson indicated the possibility of one day building a mechanical man that could talk. The automata of those days were unable to adapt to changing circumstances, but in the following century machines were built that could automatically counter disturbances to restore the desired performance of the machine. Perhaps the best known example of this is Watt's governor for the steam engine. This development led to Maxwell's 1868 paper, "On Governors," which laid the basis for both the theory of negative feedback and the study of system stability. At the same time, Bernard (1878) was drawing attention to what Cannon (1939) would later dub "homeostasis." Bernard observed that physiological processes often form circular chains of cause and effect that could counteract disturbances in such variables as body temperature, blood pressure, and glucose level in the blood.

The 19th century also saw major developments in the understanding of the brain. At an overall anatomical level, a major achievement was the understanding of localization in the cerebral cortex. Magendie and Bell had discovered that the dorsal roots of the spinal cord (the nerve bundles connected to the back of the cord) were sensory, carrying information from receptors in the body; while the ventral roots (on the belly side) were motor, carrying commands to the muscles. Fritsch and Hitzig, and then Ferrier, extended this principle to the brain proper, showing that the rear of the brain contains the primary receiving areas for vision, hearing, and touch, while motor cortex is located in front of the central fissure. All this understanding of localization in the cerebral cortex led to the 19th century neurological doctrine of connectionism, perhaps best exemplified in Lichtheim, 1885, which saw different mental "faculties" as localized in different regions of the brain, so that neurological deficits were to be explained in terms of lesions either of specific such regions, or of specific pathways linking two such regions [see Arbib,

Caplan, and Marshall (1982) for a review]. We may also note a major precursor of "modern" connectionism: the associationist psychology of Alexander Bain (1868), who represented associations of ideas by the strengths of connections between neurons that represented those ideas.

Around 1900, two major steps were taken in revealing the finer details of the brain. In Spain, Ramon y Cajal gave us exquisite anatomical studies of many regions of the brain, revealing its structure as a network of neurons. In England, the physiological studies of Charles Sherrington of reflex behavior gave us our first understanding of synapses, the junction points between the neurons. Somewhat later, Ivan Pavlov, extending associationist psychology and building on the Russian studies of reflexes by Sechenov in the 1860s, established the basic facts on the modifiability of reflexes by conditioning [see Fearing (1930) for a historical review].

Another major set of materials setting the scene for cybernetics came from work in mathematical logic in the 1930s. Kurt Gödel published his famous Incompleteness Theorem in 1931 (cf. Chapter 8). He showed that if one used the approach offered by Whitehead and Russell in their *Principia Mathematica* to set up consistent axioms for arithmetic and prove theorems by logical deduction from them, the theory *must* be incomplete—there would be true statements of arithmetic that could not be proved.

Following Gödel's 1931 study, many mathematical logicians sought to formalize the notion of an effective procedure, of what could be done by explicitly following an algorithm or set of rules. Kleene developed the theory of recursive functions, Turing his A-machines, Church developed the lambda calculus (which provided the basis for McCarthy's list processing language, LISP, a favorite of AI workers), while Emil Post introduced systems for rewriting strings of symbols (of which Chomsky's grammars are a special case). Fortunately, these methods proved to be equivalent. Whatever could be computed by one of these methods could be computed by any other method if it were equipped with a suitable "program." In particular, it came to be believed (Church's thesis) that if a function could be computed by any machine at all, it could be computed by each one of these methods. In particular, Turing (1936) helped chart the limits of the computable with his notion of what is now called a Turing machine (cf. Chapter 6), a device that could read, write, and move upon an indefinitely extendible tape, each square of which bore a symbol from some finite alphabet. As one of the ingredients of Church's thesis, he made plausible the claim that any effectively definable computation, that is, anything that a human could do in the way of symbolic manipulation by following a finite and completely explicit set of rules could be carried out by such a machine equipped with a suitable program.

During World War II, Wiener in the United States and Kolmogorov in the Soviet Union developed the theory of estimation and prediction as a contribution to the mathematical analysis of control of antiaircraft artillery. With his colleagues in England, Turing helped develop electronic computers for code-breaking; while in the United States, electronic computers were developed for the complex computations involved in many military applications, including the design of the atomic bomb. The war also saw the development of operations research, the mathematical analysis of the allocation of resources.

And so we come to 1943, the key year for bringing together the notions of control mechanism and intelligent automata:

Craik (1943) published his seminal essay *The Nature of Explanation*. Here the nervous system was viewed "as a calculating machine capable of modeling or paralleling external events," suggesting that the process of forming an "internal model" that paralleled the world is the basic feature of thought and explanation.

Rosenblueth, Wiener, and Bigelow (1943) published "Behavior, Purpose and Teleology." Engineers had noted that if feedback used in controlling the rudder of a ship was, for instance, too brusque, the rudder would overshoot, compensatory feedback would yield a larger overshoot in the opposite direction, and so on and so on as the system wildly oscillated. Wiener and Bigelow asked Rosenblueth if there were any corresponding pathological condition in humans and were given the example of intention tremor associated with an injured cerebellum. This evidence for feedback within the the human nervous system led the three scientists to urge that neurophysiology move beyond the Sherringtonian view of the central nervous system as a reflex device adjusting itself in response to sensory inputs. Rather, setting reference values for feedback systems could provide the basis for the analysis of the brain as a purposive system explicable only in terms of circular processes, that is, from nervous system to muscles to the external world and back again via receptors.

In the paper "A Logical Calculus of the Ideas Immanent in Nervous Activity," McCulloch and Pitts (1943) offered their formal model of the neuron as a threshold logic unit (Chapter 2), building on the neuron doctrine of Ramon y Cajal and the excitatory and inhibitory synapses of Sherrington. They used notation from the mathematical logic of Whitehead, Russell, and Carnap, but a major stimulus for their work was the Turing machine. McCulloch and Pitts provided the "physiology of the computable," demonstrating that each Turing machine program could be implemented using a finite network (with loops) of their formal neurons. Thus as electronic computers were built toward the end of World War II, it was understood that whatever they could do could be done by a network of neurons.

1.2 Cybernetics Defined and Dissolved

Section 1, then, exhibits some of the strands that were gathered in Wiener's 1948 book *Cybernetics*: *or Control and Communication in the Animal and the Machine* and in the Josiah Macy, Jr., Foundation conferences, which, from 1949 on, were referred to as *Cybernetics*: *Circular Causal and Feedback Mechanisms in Biological and Social Systems*.

At the Hixon Symposium of 1948 (published in 1951), John von Neumann reflected on the influence of the work of McCulloch and Pitts on the design of digital computers, and speculated on the extension of Turing's notion of a universal computer to that of a universal constructor, thus laying the grounds for the theory of self-reproducing automata (Chapter 7). At the same Symposium, Karl Lashley analyzed the problem of serial order in behavior, arguing that such serial order could not be explained by the usual approach of stimulus–response probabilities central to behaviorist psychology. If McCulloch and Pitts had shown how to build a "nervous system" for a Turing machine, then Lashley may be seen as playing a key role in bringing the notion of "stored program" into psychology. His paper is thus seminal both for the development of artificial intelligence and cognitive psychology and for the fruitful interaction between them (of which more below).

In 1949, Hebb published *The Organization of Behavior*, which provided the inspiration for many computational models of learning and adaptation in neural systems. He argued that learning was to be explained by the formation of cell assemblies in the brain. The work of Lashley and Hebb was crucial for the rising respectability of intervening variables in psychology, and the decay of the behaviorist paradigm. This paradigm was based on an attempt to free psychology of "mystical" mental constructs; the growth of cybernetics and computer programming provided a new vocabulary in which intervening variables could be given an objective description.

In 1950, Turing published the Turing test, suggesting that a computer could be considered intelligent if a human communicating by teletype failed to distinguish the machine from another person on the basis of their response to arbitrary questioning. In 1953, Merton advanced the cybernetic program by his "speculations on the servo-control of movement," which established feedback as part of the vocabulary of neurophysiologists studying the control of movement.

Hebb offered a learning scheme for units like the formal neurons of McCulloch and Pitts in which the connection between two neurons was strengthened if both neurons fired at the same time. Such a scheme tends to sharpen up a neuron's predisposition "without a teacher," getting its firing to become better and better correlated with a cluster of stimulus patterns. Thus the final set of input weights to the neuron depends both on the initial setting of the weights and on the pattern of clustering of the set of stimuli to which it is exposed. By contrast, the perceptron scheme introduced in the mid-1950s (cf. Rosenblatt, 1962) looked at various neural networks in which certain of the neurons could be modified depending on whether or not the firing of the neuron matched some pattern determined by a "teacher." The best known perceptron learning rule strengthens an active synapse if the efferent neuron failed to fire when it should have fired, and weakens an active synapse if the neuron fires when it should not have done so. In Section 4.2 we shall prove a convergence theorem for a perceptron solving a "linearly separable" pattern recognition problem; while Section 4.3 will trace the use of Hebb-type synapses (improved with an additional rule to stop all synapses "saturating") to model the development of preprocessing elements in visual cortex.

Ross Ashby's 1954 *Design for a Brain* melded observations from the psychiatric clinic and the asylum with a cybernetic account of brain function and homeostasis. Then in 1956, Shannon and McCarthy edited a collection called *Automata Studies* with contributions by von Neumann, Kleene, Minsky, MacKay, Ashby, and Uttley, among others. The contributions still seemed part of an integrated subject that might be called cybernetics. But in the same year a conference was held at Dartmouth College that introduced the term Artificial Intelligence (AI). In 1959, Newell and Simon, with their programmer Shaw, built on the earlier Logic Theorist to develop "The General Problem Solver" (GPS) as one of the first early successes of AI. Miller, Galanter, and Pribram's 1960 *Plans and the Structure of Behavior* then introduced the concepts of GPS to cognitive psychology. Minsky's 1961 review article, "Steps toward Artificial Intelligence," while still discussing the study of adaptive neural-like nets, proclaimed the emergence of AI as a separate field with the emphasis on sequential symbolic processes. From then on, many workers abandoned the theme of human–machine comparison to pursue AI as an end in itself—programming computers to exhibit some aspects of human intelligence, but with the emphasis on achieving some benchmark of performance rather than on capturing the mechanisms by which humans were themselves intelligent. Some workers tried to use concepts from AI to model human cognition using computer programs, but in developing this new form of cognitive psychology they paid scant attention to how brain mechanisms might underlie animal or human behavior.

Much work in cybernetics now deals with control problems in diverse fields of engineering, economic, and the social sciences (see, for example, the many papers published in the *IEEE Transactions on Systems, Man, and Cybernetics*), whereas the broad field of computer science has become a discipline in its own right. Here we briefly cite four subdisciplines that have crystallized from the earlier concern with the intergrated study of mind, brain, and machine.

Biological control theory. The techniques of control theory, especially the use of linear approximations, feedback, and stability analysis, are widely applied to the analysis of diverse physiological systems such as the stretch reflex, thermoregulation, and the control of the pupil.

Neural modeling. The Hodgkin–Huxley analysis of the action potential, Rall's models of dendritic function, analysis of lateral inhibition in the retina, and the analysis of rhythm-generating networks are examples of successful mathematical studies of single neurons, or of small or highly regular networks of neurons, that have developed in fruitful interaction with microelectrode studies.

Artificial Intelligence. This is a branch of computer science devoted to the study of techniques for constructing programs enabling computers to exhibit aspects of intelligent behavior, such as playing checkers, solving logical puzzles, or understanding restricted portions of a natural language such as English. Although some practitioners of AI look solely for contributions to technology, there are many who see their field as intimately related to cognitive psychology.

Cognitive psychology. The concepts of cybernetics gave rise to a new form of cognitive psychology that sought to explain human perception and problem solving not in neural terms but rather in some intermediate level of information-processing constructs. Recent years have seen strong interaction between AI and cognitive psychology.

Because cybernetics extends far beyond the analysis of brain and machine, the term *brain theory* has been introduced to denote an approach to brain study that seeks to bridge the gap between studies of behavior and overall function (AI and cognitive psychology) and the study of physiologically and anatomically well-defined neural nets (biological control theory and neural modeling). Workers in brain theory have always been concerned with the parallel activity of the many neurons in a neural circuit. In 1947, Pitts and McCulloch published "How we know universals: the perception of visual and auditory forms" in which they envisaged the visual brain as providing layers of features as a basis for pattern recognition. Their work inspired Lettvin, Maturana, McCulloch, and Pitts, 1959, to analyze "what the frog's eye tells the frog's brain" discovering the "feature detectors" in the frog retina—extending Barlow's (1953) discovery of the "bug detector"—and their layered projection in the frog tectum. This result is akin to the Hubel–Wiesel discovery of simple cells, complex cells, and hypercomplex cells in cat, and then monkey, visual cortex. Between them, they established the notion of a feature detector, a neuron that may serve as the unit of perceptual processing. All this will be discussed in Section 4.1. Selfridge (1959) developed the Pandemonium model of pattern recognition, in which a "chief demon" would listen to the "shouts" of "lesser demons," each specialized to emit a "howl" expressing their degree of confidence in the particular pattern they were specialized to recognize. By contrast, Kilmer, McCulloch, and Blum (1969) offered a network in which each module (itself a neural network), on the basis of its own input sample, formed initial estimates of the likelihood of each of a finite set of modes, and then communicated back and forth to achieve a consensus as to which mode was most appropriate for the given sample. This style of decision-making, involving communication among a net of modules rather than the imposition of the decision by a global controller may be called *cooperative computation.*

However, workers in both AI and cognitive psychology were so dominated by the metaphor "the mind is a computer" that many argued that the mind must share with the computers of the 1960's the property of being serial, of executing a series of operations one at a time. Whereas in the 1950s there was no distinction between AI and neural modeling, in the first anthology of work in AI, *Computers and Thought* (Feigenbaum and Feldman, 1963), the editors explicitly excluded papers on neural networks. They did include Samuel's "Study of machine learning using the game of checkers," but the study of learning, as well as the study of neural nets, thereafter played little role in the development of AI until the later 1970s.

It is often argued that the book by Minsky and Papert, *Perceptrons*: *An Essay in Computational Geometry* (see Section 4.4), played an important role in establishing the serial paradigm for AI, by showing the limitation of perceptrons with only one adaptive element, fed by a single layer of pre-processing elements (Section 4.2). However, I think this is to confuse the symptom with the cause. As already noted in Arbib (1964), the puffery of perceptrons in the popular press around 1960 (very similar to the current treatment of AI in general, the unbounded adulation of expert systems, and an article on Hopfield nets on the front page of the *LA Times*) had led to a reaction against Rosenblatt's work. It was in part this reaction, and in part a desire to proclaim the serial AI paradigm (of whose power they, along with other influential scientists like Newell and Simon, were by this time already more than convinced), that led Minsky and Papert to write their book. As new workers entered the AI field they may well have seen this book as a powerful argument for serial computation, even though the reader of Section 4.4 will see that their theorems do not prove the power of serial systems, but simply show that parallel networks of limited complexity are limited in their capa-bilities. However, as I have said, the conviction that intelligent machines *must* be serial preceded the publication of the book. Since I always looked for fruitful interaction between AI and brain theory, I remained unswayed by this argument. At a Symposium on Parallel Processing for Artificial Intelligence at New York University in January of 1975, I spoke of "The Implications of Parallel Processing in the Brain" in direct debate with Minsky's espousal of the essentially serial nature of intelligence. Yet by the following meeting of IJCAI, the International Joint Conference on Artificial Intelligence, Minsky was expounding the "society theory of minds" (finally published as Minsky 1986), very much in the McCulloch-based tradition of cooperative computa-tion, and one symptom of the rapprochement to be charted in Section 1.3.

Let me close this section with a brief statement about the status of brain theory (a subject discussed at greater length in *The Metaphorical Brain*). There are now models of various aspects of vision, learning, and the control of movement, with different models addressed to date on the differences in brain structure and function—for example, there are distinct models for visuo-motor coordination in monkey, frog, and fly. Other models seek to relate the function of specific brain regions, such as cerebellum, hippocampus, retina, and tectum, to details of the underlying circuitry. What is to be stressed is that in all these areas and more there is an increasing trend away from the development of ad hoc models, and toward the cumulative development of models through the interchange of concepts among theorists, and through critical but constructive interplay among theorists and experimentalist neuro-scientists. This constitutes a broad and lively field for which the term "brain theory" or "computational neuroscience" may properly be applied. However, it is not only the success of specific models that is responsible for the increasing attention being paid to theory by experimentalists in neuroscience: When the LINK was introduced in 1963 as the first laboratory computer for neuroscientists, many experimentalists resisted its use, believing that the machine would deprive them of the delicate "touch" needed for a successful experiment. However, by the mid-1970s, most experimentalists had learned to live with the computer as an experimental tool, finding that its use on-line allowed them to gather a wealth and subtlety of data that would be impossible without it. Nonetheless, the majority still distrusted the use of the computer as a tool for simulation or theory. It is perhaps as much a fruit of increasing familiarity with computers and the current vogue for artificial intelligence as it is the success of any particular brain model *per se* that is responsible for the increasing acceptance of brain theory by the community of neuroscientists.

1.3 The New Rapprochement

In Section 1.2, we noted a number of disciplines that emerged from the research on mind, brain, and machine which for a while were gathered under the banner of cybernetics. Here we should add another subject, but one whose formal study of the mind is rooted in the mathematical logic of the 1930s and 1940s without intervention of "the concepts of 1943," namely, the approach to *linguistics* known as generative grammar. Chomsky (1957) introduced this approach to the syntax of English in a way that may be seen as a specialization of the production systems introduced by Emil Post as part of the flurry of work on effective procedures. Post showed how to generate a language, in the formal sense of a set of strings of symbols, starting from a small given set of strings, and then forming new strings by applying various rewriting rules. For Post, the motivation was to derive theorems from axioms using rules of inference. Chomsky's innovation was to use such a framework to derive grammatical sentences of a natural language from a single symbol S denoting a sentence *in abstracto* using a grammar expressed as a finite set of rewriting rules. Chomsky's work has come to define perhaps the dominant school of research in linguistics [see van Riemsdijk and Williams (1986) for a masterly text on the evolution of this theory and its current status; Moll, Arbib, and Kfoury (1987) provides a briefer introduction]. However, where Chomsky's theory has focused on the specification of constraints on grammatical rules that distinguish natural languages from other sets of strings of symbols, workers in AI and cognitive psychology have been more concerned with models of *performance*, of the process whereby given a string of symbols we come to understand it, or whereby given an idea to express we produce an appropriate sentence. For many years, these two approaches to language were at loggerheads, with workers in each camp dismissing as irrelevant the research of the other. Nonetheless, in the late 1970s, there came to be increasing interaction leading to a rapprochement of AI, cognitive psychology, and linguistics, which became known as *cognitive science*.

Initially, cognitive science was dominated by the serial processing paradigm, and little attention was paid to parallel processing in general or to brain theory in particular. More recently, an increasing number of workers in cognitive science have broken away from the serial paradigm, and there has emerged a broad subject that embraces neural modeling and brain theory together with parts of AI, cognitive psychology, and linguistics. Within AI we may see "connectionism" as one component and machine vision as another. How did this come about? As the 1960's became the 1970's, the dominance of AI by the serial paradigm seemed almost complete. Thus my 1972 book *The Metaphorical Brain* was going very much against the grain when it adopted the subtitle "Cybernetics as Artificial Intelligence and Brain Theory," for it asserted that cybernetics had not been superseded by AI, that the theoretical study of brain mechanisms remained a viable discipline, and that there was much to be gained by the interchange of ideas between AI and brain theory. Its perspective of cooperative computation—involving the interaction of concurrently active systems, many of which are structured as layers of neuronlike elements—has now established itself (though, alas, I shall trace a historical process that does not ascribe a large causal role to my book) not only as the proper way to think about the brain, but also as an increasingly promising approach to the construction of the next generation of computers. I note three main reasons for this.

1. The technological developments that have focused increasing attention on parallelism and distributed computation. With the development of VLSI (the technique of very large scale integration that allows circuits with hundreds of thousands of components to be "printed" on a single chip) and computer networking (linking many computers together via an electronic network), many computer scientists are abandoning the serial computer paradigm and are seeking to understand how problem-solving can be distributed across a network of interacting highly sophisticated, concurrently active processors.

2. The fruitful interaction between workers in AI and brain theory in a few subjects such as vision. Many workers in machine vision began to use "brainlike" concepts such as computation in layered arrays of processors, even if few were prepared to become brain theorists in the sense of seeking to study and explain animal behavior or human cognition in terms of detailed facts from neuroanatomy and neurophysiology. My own group and my colleagues (at the University of Massachusetts at Amherst for the past 16 years, and now at the University of Southern California) have demonstrated the fruitful interplay between brain modeling and computer programming in such areas as vision, visuomotor coordination, learning, and linguistics. Workers in machine vision have contributed much to the understanding of the importance of parallelism in the "low-level" processing of images and of distributed problem-solving in the "high-level" visual problem of interpreting the objects in a scene and the relationships among

them. There has also been healthy interaction among workers designing robots, students of human movement, and neuroscientists analyzing brain mechanisms for the control of animal movement.

3. The recognition that large search spaces implied that sequential approaches to AI imposed practical limitations, coupled with the reemergence of learning as a viable subfield in AI, after an unaccountably long period of neglect. "Connectionism" combines the insight of studies of "learning networks" made in the 1950's and 1960's with the sophistication gained from more than two decades of studies in AI (even those, ironically, based on hostile rejection of these classic studies of learning networks, or "self-organizing systems" as they were also called). (A further analysis of the emergence of connectionism from serial AI will be given in Section 5.1.)

One interesting application of the theory of learning networks is that it may be too hard to explicitly program the behavior that ones sees in a black box, but that one may be able to drive a network by the actual input/output behavior of that box, or by some description of its trajectories, to cause it to adapt itself into a network with (approximately) that given behavior. However, as McCulloch, Arbib, and Cowan (1962) observed during the earlier heyday of self-organizing systems, "If you want a sweetheart in the Spring, don't get an amoeba and wait for it to evolve"—the point being that a learning algorithm may not solve a problem within a reasonable period of time unless the initial structure of the network is suitable. In this respect, it is interesting to note the studies by Bienenstock, Cooper, and Munro (1982) on formation of feature detectors in visual cortex (Section 4.3) and the work of von der Malsburg and Willshaw (1977) on the formation of maps between retina and tectum (or between any two other retinotopic arrays), in both of which a key concern is the interaction of synaptic adaptation with genetically predetermined connection patterns or biases. These last examples are from brain theory, rather than the connectionist literature, but their relevance to the present discussion makes it clear that little is to be gained by trying to erect sharp boundaries between the subjects. Quite the contrary. While their different goals (building an efficient machine versus understanding actual brains) must make for diversities between AI and brain theory, there are many unities too, and these become increasingly clear as AI shifts away from serial to cooperative computation (Arbib, 1975). For many connectionists, it is the unities that are at the focus: the attempt to build up a systematic understanding of what highly parallel networks can do and, where appropriate, how they can adapt to do different things. Thus the connectionist, even if he does not seek to model actual neurons, does seek an understanding of how concepts might be stored in the style of the brain, yielding an understanding of psychological or technological issues in a way that might not be achieved by writing LISP programs for a serial computer.

To close this section, let me note that the development of connectionism raises two issues with respect to computer hardware:

1. What hardware will enable more efficient simulation of parallel networks?
2. Will connectionism lead to the design of new computers of general utility?

To point 1, it is already clear that the simulation of parallel networks, whether or not they are adaptive, consumes a huge number of cycles. Thus, array processors and supercomputers that use pipelines for ultrafast handling of vectors are already being pressed into service. The connection machine developed by Thinking Machines Incorporated will also find application here. Hillis (1985) titles his report "The connection machine: Computer architecture for the new wave." Meanwhile, people are beginning to develop special purpose chips both for the highly parallel algorithms of fixed architecture such as those used in low-level vision and for networks in which adaptation can be automatically conducted with immense rapidity at the finest level of circuitry.

To point 2, it seems clear that there are many tasks for which the pattern-recognition and associative memory capability of a connectionist net will be valuable, while for other applications (e.g., a data base for insurance records or for airline reservations) a conventional bit-by-bit storage system may be preferable. A content-addressable memory can solve many data-base problems. In some cases, our increasing understanding of parallelism will suggest new ways in which data are to be explicitly entered; in other cases, it will be appropriate to build adaptive networks whereby answers may be "inferred" by some learning rule. Thus one may still want to use an associative architecture, but not an adaptive one—a learning net being too quirky and open to making false associations. Thus, we will see increasing interaction between AI and brain theory in understanding how to match the architecture of "neuronlike" systems to the problems that they are to solve. In some cases, we will seek fixed algorithms for cooperative computation to solve a problem; in other cases, we will stress the use of learning algorithms to adapt a precursor network to a poorly understood task. In between, there will be cases in which it will be valuable to use adaptation to find the weights in a network, but then to build a chip with those weights fixed in it to yield a fixed special-purpose system: for once you have something satisfactory, you may not want to change it.

References for Chapter 1

Arbib, M.A., 1964, *Brains, Machines, and Mathematics*, McGraw-Hill.

Arbib, M.A., 1972, *The Metaphorical Brain: An Introduction to Cybernetics as Artificial Intelligence and Brain Theory*, Wiley Interscience.

Arbib, M.A., 1975, Artificial intelligence and brain theory: Unities and diversities, *Ann. Biomed. Eng.* **3**: 238–274.

Arbib, M.A., 1988, *The Metaphorical Brain: An Introduction to Schemas and Brain Theory*, Wiley Interscience.

Arbib, M.A., Caplan, D., and Marshall, J.C., 1982, Neurolinguistics in historical perspective, in *Neural Models of Language Processes* (M.A. Arbib, D. Caplan, and J.C. Marshall, Eds.), Academic Press, pp. 5–24.

Ashby, W.R., 1954, *Design for a Brain*, Chapman and Hall.

Bain, A., 1868., *The Senses and the Intellect*, Third edition.

Barlow, H.B., 1953, Summation and inhibition in the frog's retina, *J. Physiol. (Lond.)* **119**: 69–88.

Bernard, C., 1878, *Lecons sur les phénomènes de la Vie*, Bailliére.

E.L. Bienenstock, L.N. Cooper, and P.W. Munro, 1982, Theory for the development of neuron selectivity; Orientation specificity and binocular interaction in visual cortex, *J. Neuroscience* **2**: 32–48.

Cannon, W.B., 1939, *The Wisdom of the Body*, Norton.

Chomsky, N., 1957, *Syntactic structures*, Norton.

Craik, K.J.W., 1943, *The Nature of Explanation*, Cambridge University Press.

Fearing, F., 1930, *Reflex Action*, Williams and Wilkins.

Feigenbaum, E., and Feldman, J., Eds., 1963, *Computers and Thought*, McGraw-Hill.

Gödel, K., Uber Formal unentscheidbare Sätze der Principia Mathematica und verwandter Systeme, I, *Monats. Math. Phys.* **38**: 173–198.

Hebb, D.O., 1949, *The Organization of Behavior*, John Wiley & Sons.

Hillis, W.D., 1985, *The Connection Machine*, The MIT Press.

Jeffress, L., Ed., 1951, *Cerebral Mechanisms in Behavior*, Interscience.

Kilmer, W.L., McCulloch, W.S., and Blum, J., 1969, A model of the vertebrate central command system, *Int. J. Man-Machine Studies* **1**: 279–309.

La Mettrie, J., 1953, *Man a Machine* (translated by G. Bussey), Open Court.

Lashley, K.S., 1951, The problem of serial order in behavior, in *Cerebral Mechanisms in Behavior: The Hixon Symposium* (L. Jeffress, Ed.), Wiley, pp. 112–136.

Lettvin, J.Y., Maturana, H., McCulloch, W.S., and Pitts, W.H., 1959, What the frog's eye tells the frog brain, *Proc. IRE.* 1940–1951.

Lichtheim, L., 1885, On aphasia, *Brain* **7**: 433–484.

Maxwell, J.C., 1868, On governors, *Proc. R. Soc. London* **16**: 270–283.

McCulloch, W.S., Arbib, M.A., and Cowan, J.D., Neurological models and integrative processes, in *Self-Organizing Systems, 1962* (M.C. Yovits, G.T. Jacobi and G.D. Goldstein, Eds.), Spartan Books, pp. 49–59.

McCulloch, W.S., and Pitts, W.H., 1943, A logical calculus of the ideas immanent in nervous activity, *Bull. Math. Biophys.* **5**: 115–133.

Merton, P.A., 1953, Speculations on the servo-control of movement, in *The Spinal Cord: A CIBA Foundation Symposium*, J. and A. Churchill Ltd., pp. 247–255 (Discussion, pp. 255–260).

Miller, G.A., Galanter, E., and Pribram, K.H., 1960, *Plans and the Structure of Behavior*, Henry Holt & Co.

Minsky, M.L. 1961, Steps toward artificial intelligence, *Proc. IRE* **49**: 8–30.

Minsky, M.L., 1986, *The Society of Mind*, Simon & Schuster.

Minsky, M.L., and Papert, S., 1969, *Perceptrons: An Essay in Computational Geometry*, The MIT Press.

Moll, R.N., Arbib, M.A., and Kfoury, A.J., 1987, *An Introduction to Formal Language Theory*, Springer-Verlag.

Newell, A., Shaw, J.C., and Simon, H.A., 1959, Report on a general problem-solving program, *Proc. Int. Conf. Info. Processing*, UNESCO House, pp. 256–264.

Pitts, W.H., and McCulloch, W.S., 1947, How we know universals: The perception of auditory and visual forms, *Bull. Math. Biophys.* **9**: 127–147.

Rosenblatt, F., 1962, *Principles of Neurodynamics*, Spartan Books.

Rosenblueth, A., Wiener, N., and Bigelow, J., 1943, Behavior, purpose and teleology, *Philos. Sci.* **10**: 18–24.

Samuel, A., 1959, Some studies in machine learning using the game of checkers, *IBM. J. Res. and Dev.* **3**: 210–229.

Selfridge, O.G., 1959, Pandemonium: A paradigm for learning, in *Mechanisation of Thought Processes*, Her Majesty's Stationery Office, pp. 511–531.

Shannon, C.E., and McCarthy, J., Eds., 1956, *Automata Studies*, Princeton University Press.

Turing, A.M., 1936, On computable numbers with an application to the Entscheidungsproblem, *Proc. London Math. Soc. Ser. 2* **42**: 230–265.

Turing, A.M., 1950, Computing machinery and intelligence, *Mind* **59**: 433–460.

van Riemsdijk, H., and Williams, E., 1986, *An Introduction to the Theory of Grammar*, The MIT Press.

von Neumann, J., 1951, The general and logical theory of automata, in *Cerebral Mechanisms in Behavior: The Hixon Symposium* (L.A. Jeffress, Ed.), Wiley, pp. 1–32.

von der Malsburg, C., and Willshaw, D.J., 1977, How to label nerve cells so that they can interconnect in an orderly fashion, *Proc. Natl. Acad. Sci. USA* **74**: 5176–5178.

Wiener, N., 1948, *Cybernetics: or Control and Communication in the Animal and the Machine*, The Technology Press and Wiley (Second Edition, The MIT Press, 1961).

CHAPTER 2

Neural Nets and Finite Automata

2.1 Logical Models of Neural Networks

I want to start by giving a very sketchy account of neurophysiology—merely sufficient as a basis for our first mathematical model. We may regard the human nervous system as a three-stage system as shown in Figure 2.1.*

Our fundamental hypothesis in setting up our model is that all the functioning of the nervous system relevant to our study is mediated solely by the passage of electrical impulses by cells we call neurons. Actually, the human brain contains more *glial* cells than it contains *neurons*, but it is neurophysiological orthodoxy to believe that these glial cells served only to support and nourish the neurons—functions irrelevant to our study. Throughout this book, we shall ignore such posited glial functions. We shall also ignore such modes of neural interaction as continuously variable potentials and transmission of hormones. In setting up our *possible* mechanisms, neural impulses will fully suffice— future developments will, of course, require the ascription of far greater importance to the other neural functions and perhaps to the glia.

In the light of our fundamental hypothesis, then, we shall simply view the nervous system proper as a vast network of neurons, arranged in elaborate structures with extremely complex interconnections. This network receives inputs from a vast number of *receptors*: the rods and cones of the eyes; the pain, touch, hot, and cold receptors of the skin; the stretch receptors of muscles; and so on; all converting stimuli from the body or the external world

---

*The purpose of the arrows drawn from right to left will be made clear in the discussion of feedback in Section 3.1

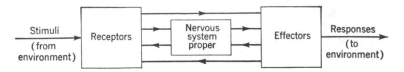

Figure 2.1   The nervous system considered as a three-stage system.

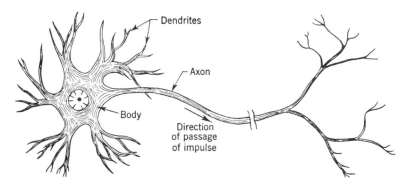

Figure 2.2   Schematic drawing of a neuron.

into patterns of electrical impulses that convey information into the network. These interact with the enormously complicated patterns already traveling through the *neural net* (there are estimated to be more than 10^{10} neurons in the neural net which is the human brain) and result in the emission of impulses that control the *effectors*, such as our muscles and glands, to give our responses. Thus we have our three-stage system: receptors, neural net, and effectors.

We are not going to formulate models of the receptors or effectors here, but we do want a model of the neural net. To do this, we first model the neuron. The neurons of our nervous system come in many forms, but we shall restrict our study to neurons like that of Figure 2.2.

The *neuron* is a cell and so has a nucleus, which is contained in the *soma* or *body* of the cell. One may think of the *dendrites* as forming a very fine filamentary bush, each fiber being thinner than the axon, and of the *axon* itself as a long, thin cylinder carrying impulses from the soma to other cells. The axon splits into a fine arborization, each branch of which finally terminates in a little *endbulb* almost touching the dendrites or other parts of a neuron. Such a place of near contact is called a *synapse*. Impulses reaching a synapse set up graded electrical signals in the dendrites of the neuron on which the synapse impinges, the interneuronal transmission being sometimes electrical but usually by diffusion of chemicals called *transmitters*. A particular neuron will only fire an electrical impulse along its axon if sufficient impulses reach the endbulbs impinging on its dendrites in a short period of time, called the *period of latent summation.* Actually, these impulses may either help or hinder the firing of an impulse and are correspondingly called *excitatory* or *inhibitory*. The condition for the firing of a neuron is then that the excitation should exceed the inhibition by a critical amount called the *threshold of the neuron*. If we assign a suitable positive weight to each excitatory synapse and a negative weight to each inhibitory synapse, we can say that

  a neuron fires only if the total weight of the synapses that receive impulses in the period of latent summation exceeds the threshold.

There is a small time delay between a period of latent summation and the passage of the corresponding axonal impulse to its endbulbs, so that the arrival of impulses on the dendrites of a neuron determines the firing of its axon at a slightly later time. After an impulse has traveled along an axon, there is a time called the *refractory period* during which the axon is incapable of transmitting an impulse. We can thus convey some of the information about the train of impulses coursing along the axon of a neuron by dividing our time scale into consecutive intervals, each of length equal to one refractory period of the given neuron, and then specifying for each such interval whether or not a spike was generated in that interval. We might then hope to specify of a neuron whether or not it will "fire" in period $n + 1$ by specifying which of the incoming axons "fire" in period n. In other words, we approximate the neuron by a "binary switching device" that can be switched on or off in successive intervals of time as a function of the states of its inputs in the previous interval. In modeling real neurons, we may consider the time unit to be on the order of a thousandth of a second.

These highly simplified considerations lead to the model of the neuron introduced by Warren McCulloch and Walter Pitts, 1943. Let $x_i(n) = 1$ if the ith of the input lines of our formalized neuron bears a pulse during the nth time interval on our time scale; let it equal 0 if not. If the ith synapse is *excitatory*, we associate with it a number w_i greater than 0, representing the amount of transmitter substance that a pulse would release at the synapse. Similarly, w_i *less* than 0 would represent an *inhibitory* synapse [Figure 2.3(a)].

1 Definition. A *McCulloch–Pitts neuron* (or *threshold logic unit*) is an element with, say, m inputs $x_1, \ldots, x_m$ ($m \geq 1$) and one output y. It is characterized by $m + 1$ numbers, its threshold θ, and the weights $w_1, \ldots, w_m$, where w_i is associated with x_i. Taking a refractory period as the unit of time, we postulate that the neuron operates on a time scale $n = 1, 2, 3, 4, \ldots$, the firing of its output at time $n + 1$ being determined by the firing of its inputs at time n according to the following rule: the neuron fires an impulse along its axon at time $n + 1$ if the weighted sum of its inputs at time n exceeds the threshold of the neuron, which rule may be expressed symbolically as

$$y(n + 1) = 1 \quad \text{if and only if} \quad \sum_i x_i(n) \geq \theta.$$

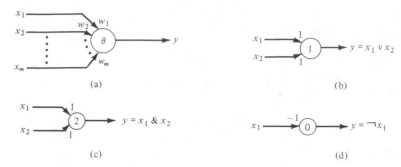

(a)

(b)

(c)

(d)

Figure 2.3   The general scheme (a) for a McCulloch–Pitts neuron (threshold logic unit) showing the weights w_1, w_2, ..., w_m and threshold θ. (b) The OR gate: $x_1 + x_2 \geq 1$ so long as either x_1 or x_2 is 1; $x_1 \vee x_2$ (vel is Latin for this or) denotes "x_1 OR x_2." (c) The AND gate: $x_1 + x_2 \geq 2$ only if both x_1 and x_2 are 1; $x_1 \& x_2$ denotes "x_1 AND x_2." (d) The NOT gate: the input weight is -1, corresponding to an inhibitory synapse. The total excitation of $-x_1$ can only attain threshold 0 if x_1 is itself 0. The output fires at time $t + 1$ just in case the input does *not* fire at time t; $\neg x_1$ denotes "NOT x_1."

Note that a positive weight $w_i > 0$ corresponds to an excitatory synapse, whereas a negative weight $w_i < 0$ means that x_i is an inhibitory input.

To illustrate this simple model of a neuron, we present the examples of Figure 2.3. Figures 2.3(b) and 2.3(c) only involve excitatory inputs; Fig. 2.3(d) has an inhibitory input. It is well known to computer scientists that any computer can be built just using AND, OR, and NOT gates, and the three figures show how each type of gate can be built using a single McCulloch–Pitts neuron. Thus, *highly simplified though these model neurons may be, any computer can be simulated by a network of such neurons*, and we shall spell this out in more detail below. The moral of this is that we can achieve highly complex computation with very simple components; the crucial factor is not the complexity of components, but rather the way in which they are hooked together, so that information can be carried by the pattern of activity in the neural interconnections. This simple model only preserves information about whether or not a neuron fires in any time unit, discarding information about the time between each pair of spikes (compare spike trains 1 and 2 in Figure 2.4). What it tells us is that with these simplest of neurons we can carry our arbitrary computations which demand a finite memory, so that what happens next (both in external action and in updating the memory) depends only on the current memory state and current input.

We thus put an end to any argument that says of many types of specified behavior, "That's too complicated: a bunch of neurons couldn't do that." We know that a bunch of neurons can (more of the formal theory of "what machines can do" in Chapter 6). The real question is: How is the structure and complexity of real neurons suited to carry out these functions in an efficient

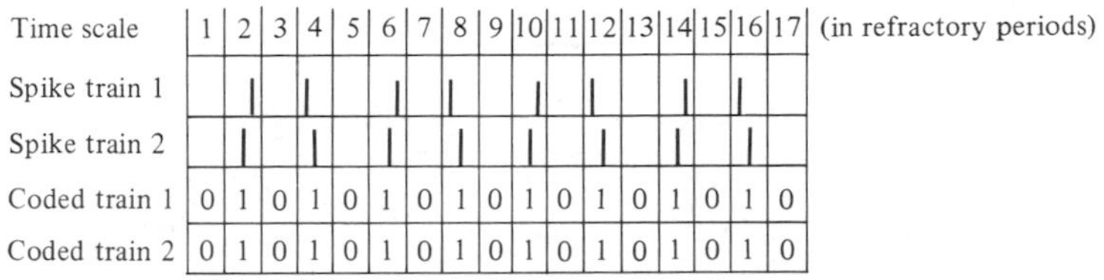

Time scale	1	2	3	4	5	6	7	8	9	10	11	12	13	14	15	16	17	(in refractory periods)
Spike train 1																		
Spike train 2																		
Coded train 1	0	1	0	1	0	1	0	1	0	1	0	1	0	1	0	1	0	
Coded train 2	0	1	0	1	0	1	0	1	0	1	0	1	0	1	0	1	0	

Figure 2.4 Two spike trains that differ in timing details even though they yield the same sequence of 0's and 1's.

and compact way? The fact is that genetically specified computing elements have evolved to do particular jobs very fast. We shall ask, "What is the natural range of computations for the organism?" and, "What type of components will allow the organism to carry out those computations very quickly?" and then, "Can we, in fact, see any relation between the components in the neural system of the animal and the components we would predict in our general theory?" That is where we would like to be, although when we come to talk about complexity of computation in Section 4.4 we shall see that at present we can only prove simple results, so that the usefulness of the theory at the moment is more qualitative in indicating new directions in which we might like to aim research, rather than quantitative in letting us actually tell the experimentalist, "Stick your electrode in here, take an electron micrograph there, and this is what you'll find."

In terms of the McCulloch–Pitts model of a neuron, we may immediately define our first model of a neural net:

2 Definition. A *neural net* is a collection of McCulloch–Pitts neurons, each with the same time scale, interconnected by splitting the output of any neuron into a number of lines and connecting some or all of these to the inputs of other neurons. An output may thus lead to any number of inputs, but an input may come from at most one output.

The *input lines* of a net are those inputs $l_0, l_1, \ldots, l_{m-1}$ of neurons of the net that are not connected to neural outputs. The *output lines* of a net are those lines $p_0, p_1, \ldots, p_{r-1}$, (branches of) neural outputs, that are not connected to neural inputs. In the example of Figure 2.5, there are three input lines and four output lines—note that the input lines may split and that the output lines need not come from distinct neurons.

We have now set up a first, highly general, model of the brain. To the reader who thinks of a model as an actual collection of wires and transistors, my use of the word "model" here may seem somewhat strange. The *engineer* feels he has modeled a system when he has actually constructed an apparatus that he can hope will behave similarly to the original system. The *computer scientist* feels he has modeled a system when he has simulated it on a computer. The *mathematician*, on the third hand, feels that he has modeled a system when he

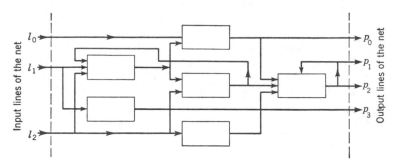

Figure 2.5   A simple neural net.

has "captured" some properties of the system in precise mathematical definitions and axioms in such a form that he can deduce further properties of this
"formal" (i.e., mathematical) *model*; thus, the mathematician hopes, *explain-
ing* known properties of the original system and *predicting* new properties. The
concept of a "neural net" has a precise mathematical definition given by
Definition 2 (and we shall prove theorems about it in subsequent sections),
and it is in this mathematical sense that we consider it to be a model of the
brain. Before we study it, let us stress that *we have only obtained it at the cost
of drastic simplifications*:

(a) We have enforced the *binary-state* and *discrete-time* assumptions: we have
     assumed that each neuron may be characterized at any time by a single 1
     or 0 value, firing or not firing; and we have assumed complete synchroni-
     zation of all the neurons, with all neurons able to change state at most
     once in each period of a common time scale, $n = 0, 1, 2, 3, \ldots$.
(b) We have fixed the threshold and weights of each neuron for all time.
(c) We have ignored the effects of hormones and chemicals (e.g., alcohol) in
     changing the behavior of the brain.
(d) We have ignored all interaction between neurons (e.g., due to the electrical
     field associated with their impulses) save that taking place at the synapses.
(e) We have ignored the glial cells.

The list can be extended, and it must be realized that *our first model is only
a starting point for our study and not an end in itself*. However, our simplica-
tions have not rendered our model completely powerless, and a neural net of
the kind specified in Definition 2 can indeed store information and carry out
computations. We shall demonstrate this in Section 2.2 by "blueprinting" a
digital computer as a neural network. First, we introduce the concept of
"finite automaton" relating it to that of "neural network" in Theorem 2.2.2.
    In Chapters 4 and 5, we shall relax restriction (b) and consider models of
learning in which elements of a neural network are allowed to change the
values of their connection weights on the basis of experience. However, more
realistic models of the brain differ from the models of this book in two ways:

1. They reject the discrete-time/binary-state assumption, modeling a neuron not in terms of a binary output changing value in terms of discrete time steps, but rather in terms of membrane potentials varying continuously in space and time.
2. They use anatomical data which show that different parts of the brain are composed of very different types of neurons. They thus seek to relate the function of each brain region to its distinctive neuron types and their specific patterns of connectivity.

Our discussion of the frog retina in Section 4.1 will give a brief taste of (2), but the range of such models is beyond the scope of the present volume, which seeks to explore the capabilities of networks of neurons obeying the binary-state/discrete-time hypothesis, thus setting a lower bound to the capabilities of real brains, composed as they are of more complex neurons. Shepherd (1979) gives a very elegant account of different regions of the brain in terms of their distinctive patterns of neural form and synaptic interaction; models of animal and human behavior in terms of interacting brain regions or functional subsystems are given in *The Metaphorical Brain* (Second Edition).

2.2 States, Automata, and Neural Nets

In the previous section, we introduced a particular type of system, a neural net as exemplified in Figure 2.5, where we can look inside the "box" to see the internal structure. However, we shall start the present section by considering a generic system as a "black box" (Figure 2.6). The environment acts upon it (it receives *inputs*) and measurements may be made of its activity (it emits *outputs*). In an assembly line, the inputs might be components and the outputs might be some assembled product; in computers, the input and output could both be information encoded in sequences of symbols; a house might be defined by money in, garbage out; while for an organism viewed as an information processor, the input might be the pattern of receptor activity elicited by environmental stimulation, while the output might be the pattern of muscular contraction controlled by motoneuron activity (compare Figure 2.6 with Figure 2.1).

A real choice is required in specifying the set of inputs and the set of outputs that are to enter into our mathematical analysis of a system. Consider our example of an organism. In an analysis of normal activity of the organism, we

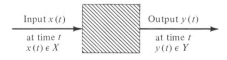

Figure 2.6   A system considered as a "black box."

might consider the input to be the visual stimulation provided by normal three-dimensional objects. If, however, we are trying to study optical illusions, then the input set must be extended to include all two-dimensional outlines and not simply those of real objects. An even further analysis might include fine details corresponding to quantum effects in the retina. Similarly, for output we might take the verbal report of a human, or detailed measurements of neural activity. Thus, our choice of the sets X of inputs and Y of outputs will depend on the range of problems we expect to cover with our model of the system. Our model only encompasses that which we believe to be relevant.

Having decided (and at some stage in analysis, we may have to revise this decision) what are the appropriate input and output sets for our model, we must now determine on what time scale we shall analyze the system. If we consider monitoring an electroencephalogram (EEG) so that the output is a set of potential differences between scalp electrodes at any time, then we may think of the output variables as changing continuously and so our model becomes a *continuous-time system* in which the time set T is indexed by the set of all real numbers. If, on the other hand, we are studying the learning curve of a rat, then we present stimuli to the system one at a time, and we note the responses of the system one at a time, and so a continuous description of the input is inappropriate, and we prefer to model the rat (in this context) as a *discrete-time system* in which the time set T is indexed by the set of all integers, in which successive integers encode the times at which successive responses are elicited from the rat. Of course, it may be appropriate to restrict our attention just to times T greater than or equal to 0 if, for instance, 0 is the time at which we start studying our system. Again as our rat example suggests, we need not think of the successive times of a discrete-time system as occurring regularly with some respect to some real time; they may correspond to successive trials, with one pair of trials being separated by 5 seconds while days may separate another pair.

For our present purposes (we shall be more specific below), an *automaton* may be regarded as a *discrete-time system with finite input and output sets*, and we shall discuss ways in which this type of abstraction may help us understand neural signal processing. Calculus provides the mathematical tools for continuous-time systems, whereas algebra provides the main tools for the automata theory, which is our central concern here. The Appendix presents the notions of set theory needed for the following exposition.

If we apply inputs to our system and monitor its corresponding outputs, can we expect the outputs to be uniquely determined by the sequence of inputs? In general, the answer is no. If we come up to a child at some time t_0 and speak to him until time t_1, we know that the responses we get during that period will not depend only on what we say, but also on whether or not the child is awake, and many other things, at time t_0. In other words, to specify how the inputs to any system yield its outputs, we need a description of some internal *state* of the system (Figure 2.7).

We must thus add to our specification of the sets T, X, and Y of time,

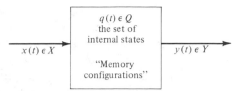

Figure 2.7   A system with an explicit set of internal states mediating the relation between input sequences and outputs.

inputs, and outputs, a set Q of internal states or "memory configurations." (The reader is warned that there are almost as many notational conventions as there are authors in automata and systems theory, and thus other books use X for the set of states, and so forth.)

We can now imagine two quite different cases. In one, having specified the state and the subsequent inputs, we can *determine* uniquely the subsequent states and outputs. In the other case, no matter how fully we specify the state, and no matter how carefully we specify subsequent inputs, we cannot specify exactly what will be the subsequent states and outputs. If we are lucky, we can give tight probabilities on the possible states and outputs; in that case, we have a system that is called *stochastic*. Here we shall only study *deterministic* systems.

Stochastic systems are, however, worthy of study. Some systems (for example, neural membranes) are studied at the level of quantum mechanics where many physicists believe that Heisenberg's uncertainty principle decrees that no matter how carefully we specify the system we can at best make probabilistic predictions about its future behavior. On the other hand, if we have a system with an immense number of variables, then even though the environment and the system are completely deterministic, we may want (as when we pass from neural patterning to gross EEG recordings) an analysis in which we only look at certain overall patterns to get the input and output sets we use in our analysis. Those variables we ignore become "noise," and can only be included in our analysis by introducing statistical perturbations of the dynamics. Thus, a stochastic treatment is worthwhile either because we are analyzing systems at the quantum level or because we are analyzing macroscopic systems that lend themselves to a stochastic description by ignoring fluctuations in microscopic variables.

Let us then be quite specific about the notion of state:

> The **state of a deterministic system** *is some representation of the past activity of the system that is sufficiently detailed to serve as a basis together with the current inputs for determining what the next output and state will be.*

For instance, if we receive a computer from the factory, "in its 0 state," and then start using it, we expect that its state will change. One way of describing how this state has changed is simply to provide a record of all input history up to the time at which we come to analyze the machine. But this incredible record of data is far too redundant, and we know that to describe the way a computer will act it is sufficient to specify for its state the present contents of its registers and memory locations. Thus, the state is some compact description, usually far less redundant than a complete input history, which allows us to predict the future response of the machine to specified inputs. At this stage, we could give a general development of the mathematical formulation of a system that embraces all the deterministic examples discussed above, but instead we turn directly to the case of interest to us, the discrete-time, time-invariant [recall assumption (b) from Section 2.1] system with a finite set of inputs and outputs:

1 Definition. An *automaton* is specified by three sets X, Y, and Q, and two function δ and β, where

  (i) X is a finite set, the set of *inputs*;
 (ii) Y is a finite set, the set of *outputs*;
(iii) Q is the set of *states*; while
 (iv) $\delta: Q \times X \to Q$, the *next-state function* is such that if at any time t the system is in state q and receives input x, then at time $t + 1$ the system will be in state $\delta(q, x)$; and
  (v) $\beta: Q \to Y$, the *output* function, is such that state q always yields output $\beta(q)$.

We say that the automaton is *finite* if Q is a finite set.

   We now want to make explicit what by now may be intuitively clear to the reader, that every neural net (in the sense of Definition 2.1.2) may be viewed as a finite automaton. More subtly, we shall then prove the converse: that the input–output behavior of a finite automaton can always be carried out by a suitably constructed neural net. Since it is much easier, in general, to design a finite automaton for a given task than to design the corresponding neural net, the result clarifies for us what tasks our neural nets are capable of performing.
   To prove our first result, let the neural net N have m input lines, q neurons, and r output lines. We say that we know the *input* of the net when we know which of the m input lines are firing and which are not firing at that time. Thus there are 2^m inputs, since we can assign the values "on" and "off" to the m input lines in 2^m different ways. Similarly, we have 2^r outputs. We say that we know the *state* of the net at time t if we know which of the q neurons are firing and which are not firing at that time. Thus there are 2^q states. We denote by Q the set of states, by X the set of inputs, and by Y the set of outputs of our net N.
   The firing of a neuron of N at time $t + 1$ is determined by the firing pattern of its inputs at time t and so, a fortiori, is determined by the state and input of the whole net N at time t. But this means that *the input and state of the net at*

time t determine the output and state of the net at time $t + 1$. Clearly, then, *any neural net is a finite automaton.*

What is more surprising is that any finite automaton can, essentially, be replaced by a neural net. The details of this replacement are not at all profound, and the reader who is not interested in them may skip the next page or so. If our finite automaton A has m possible inputs $x_0, \ldots, x_{m-1}$ and r possible outputs $y_0, \ldots, y_{r-1}$, we construct a neural net N with m *input lines* $h_0, \ldots, h_{m-1}$ (note that N thus has 2^m *inputs*) and r output lines $p_0, \ldots, p_{r-1}$. We associate the input x_j to A with the input $\bar{x}_j$ of N in which the only input line that fires is h_j. Similarly, we define an output $\bar{y}_j$ of N. Our desired net N then comprises $nr + m$ neurons, where the neuron labeled (k, j) corresponds to state k and input x_j of our automaton A, and the neuron labeled (k) corresponds to output y_k. Output line p_k of our net N is taken from the neuron (k).

We arrange our connections so that the neuron (k, j) is on at time $t + 1$ if and only if the automaton A was in the kth state and received the input x_j at time t; and the module (k) is on at time $t + 1$ if and only if the automaton A emits output y_k at time t.

Let $\{k_1, k_2, \ldots, k_{n(k)}\} = \{(x, j) | \delta(x, j) = k\}$, i.e., these are the *state-input* pairs which send the automaton A into the kth *state.*

Using the symbolism $\wedge$ for "and," $\vee$ for "or," and noting that $k_1(t)$ is to be read "k_1 fires at time t," etc., we have that the neuron (k, l) is to have the function $(k, l)(t + 1)$ if and only if $h_l(t) \wedge [k_1(t) \vee \cdots \vee k_{n(k)}(t)]$. That is, it fires at time $t + 1$ if and only if the state of A at time t is to be the kth, and the input to A at time t is the lth.

Let $\{\bar{k}_1, \bar{k}_2, \ldots, \bar{k}_{m(k)}\} = \{(x, j) | \beta(x) = k\}$, i.e., these are the state input pairs which cause the automaton A to emit the kth *output.* Then the neuron (k) is to have the function $(k)(t + 1)$ if and only if $\bar{k}_1(t) \vee \cdots \vee \bar{k}_{m(k)}(t)$. Since we have seen that $\wedge$ and $\vee$ gates can be built using McCulloch–Pitts neurons, one can extend the argument to check that each neuron can indeed be obtained by suitable choice of weights and threshold. We can see now that our net N satisfies the following theorem:

2 Theorem. *Let* $A = (X, Y, Q, \delta, \beta)$ *be any finite automaton:*

$$X = \{x_0, \ldots, x_{m-1}\},$$

$$Y = \{y_0, \ldots, y_{r-1}\},$$

$$Q = \{q_0, \ldots, q_{k-1}\}.$$

Then there exists a neural net N, *subsets* $\bar{x}_0, \ldots, \bar{x}_{m-1}$ *of its inputs,* $\bar{y}_0, \ldots,$ $\bar{y}_{r-1}$ *of its outputs, and* $\bar{q}_0, \ldots, \bar{q}_{k-1}$ *of its states such that if input sequences* $x_{j_1}, \ldots, x_{j_m}$ *to* N *initially in state* $\bar{q}_j$ *yields output sequences* $y_{k_1}, \ldots, y_{k_n}$, *then input* $\bar{x}_{j_1}, \ldots, \bar{x}_{j_n}$ *to* N *initially in state* $\bar{q}_j$ *yields output* $\bar{y}_{k_1}, \ldots, \bar{y}_{k_n}$ (*with a delay of at most one time unit*).

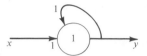

Figure 2.8   A neuron that "remembers" if it has ever received a nonzero input.

What this overelaborate piece of symbolism means is simply that the input–output behavior of a finite automaton can always be replaced by a restriction of the input–output behavior of a modular net. For many interesting examples of constructions of computing circuits from McCulloch–Pitts neurons see Minsky, 1967, Part I.

The crucial point to observe here is that for a brain, even in its simplified representation as a finite automaton, *the current output of the network need in no sense be a response to the current input regarded as stimulus.* Rather, the firing of those neurons that provide the output can be influenced by activity within the network reflecting quite ancient history of the system.

We can make this last fact very clear by an extremely simple example (see Figure 2.8). We consider a neuron with two inputs, one from outside and one taken as a branch of its own output. The weights and threshold are so chosen that the output will fire at time $n + 1$ if either input fires at time n. Suppose the neuron is initially quiet, $y = 0$. Then if the input x stays off, the neuron will stay off. But as soon as $x = 1$, it will be turned on, and a pulse will "reverberate" around the loop, keeping the neuron permanently on. Thus, the neuron "remembers" indefinitely whether its input line has ever been switched on. However, note that it can then remember nothing else.

Since any finite automaton can be replaced by a neural net, it is now easy to demonstrate the capability of a neural net for memory and computation. We do this by constructing a serial computer from interconnected finite automata—for we then know that (perhaps after a little juggling with delays) we can replace these automata by neural networks. A serial computer provided with programs (lists of instructions) of 1987 vintage is, of course, a far less "intelligent" object than a brain provided with an education of 1987 vintage—here we merely show that our initial brain model, crude as it is, at least subsumes these computers.

The last 20 years has seen the development of many ways of making more efficient hardware, including pipelining, hardwired stacks, and array processing. We also see increasing use of computer graphics and networks of concurrently active processors. These advances make for faster, more efficient computing, but do not change the class of problems that can be solved (without worrying about time constraints) using a computer. We thus consider the most basic design of a computer as made up of four units: an input–output unit, a store, a logic (i.e., logical control) unit, and an arithmetic unit. The input–output unit serves to read input instructions and data and transfer them to the store and, conversely, to transfer the results of computations from the store to the output. The store contains a finite number of "pigeonholes" each with an address, and each holding one "word"—where the "word" may be either a number, or any other data of use in computation, or an instruction. The logic unit takes one instruction at a time from the store and executes it. If it is an input–output or refer-to-memory operation, it actuates the input–output unit or store; if it is an arithmetic operation, it actuates the arithmetic unit; if it is a branch operation, it carries out a test to decide on its next instruction.

For definiteness, let us consider a computer whose instructions comprise an operation command followed by three store addresses:

$$\text{Op} \qquad \text{Operand 1} \qquad \text{Operand 2} \qquad \text{Next Instruction}$$

The first two addresses tell the logic unit where to find its operands (i.e., its data), the last tells it where to find the next instruction. For example, one might have

$$\text{Add} \qquad 3275 \qquad 3119 \qquad 4006$$

to be interpreted as: Add the numbers stored at addresses 3275 and 3119 and then execute the instruction stored at address 4006. We now describe the automata.

The state of the *store*, Figure 2.9, when it contains the words $w_1, \ldots, w_n$ in its n "pigeonholes" is simply labeled $(w_1, \ldots, w_n)$. The inputs are of the form $(m, b, 0)$ and $(m, 0, d)$. Input $(m, b, 0)$ causes the word b to be stored at address m; i.e., its state changes from to $(w_1, \ldots, w_{m-1}, w_m, w_{m+1}, \ldots, w_n)$ to $(w_1, \ldots, w_{m-1}, b, w_{m+1}, \ldots, w_n)$, but there is no output. The store has four output lines, two to the arithmetic unit, one to the input–output unit, and one to the logic unit, $a1$, $a2$, i/o, and l, respectively. Input $(m, 0, d)$ causes the word with address m to be transferred along the d-output line (where d is $a1, a2, i/o$, or l) to the appropriate unit; i.e., state $(w_1, \ldots, w_m, \ldots, w_n)$ remains unchanged and w_m is sent out along the d output. The store is clearly a finite automaton.

The *arithmetic unit* (Figure 2.10) has three input lines, one (Op) from the logic unit and two ($a1$ and $a2$) from the store. It comprises two registers, and their contents B_1 and B_2 determine the state (B_1, B_2) of the unit. The inputs are triples $(Op, a1, a2)$. Input $(0, a1, 0)$ causes $a1$ to be stored in the first register: (B_1, B_2) changes to $(a1, B_2)$ with no output. Similarly, $(0, 0, a2)$ changes (B_1, B_2) to $(B_1, a2)$. Input $(Op, 0, 0)$ causes the computation Op to be carried

Figure 2.9   The "store" automaton.

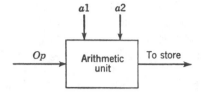

Figure 2.10   The "arithmetic-unit" automaton.

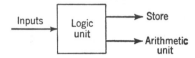

Figure 2.11   The "logic-unit" automaton.

out on (B_1, B_2) to form the result C, some remnant of the calculation D also remaining, to give state (C, D). The input $(St, m, 0)$ causes the arithmetic unit to store the result of the computation in location m of the store; i.e., the state (C, D) remains unchanged, while the unit outputs $(m, C, 0)$ to the store. Again, it is clear that this unit is a finite automaton—it has a specified response for each of its finite number of input-state pairs.

The input to the logic unit (Figure 2.11) is simply an instruction from the store. Its input line is connected to output line l of the store. Input $Op\ a\ b\ c$ causes it to change state to (Op, a, b, c). If Op is the name of a function, the logic unit then emits four outputs, one after another, in the following order: $(a, 0, a1)$ and then $(b, 0, a2)$ to the store, which as a consequence, transmits the operands to the arithmetic unit; the logic unit then sends the command $(Op, 0, 0)$ to the arithmetic unit, thus causing the operation to be carried out on those operands; and, fourth, it sends $(c, 0, l)$ to the store, causing it to prime the logic unit with the next instruction.

If Op is a branch operation Br (so that $Br\ a\ b\ c$ means test word at address a; if the answer is "yes," next instruction is at b; if "no," it is at c), one can devise the logic unit to carry out the test in a finite way and accordingly emit $(b, 0, l)$ or $(c, 0, l)$ to the store. Anyway, the logic unit is a finite automaton. These details suffice for our construction of a computer. I leave the construction of an *input/output* finite automaton (and the construction of corresponding inputs for the logic unit) as an exercise for the reader.

Those readers who have had any experience with computers will know that a digital computer is made up of a network of hundreds of thousands of transistorlike elements integrated onto VLSI "chips," and in this light it is of course obvious that a computer is a "neural net," even though the "neurons" are not strictly our McCulloch–Pitts neurons. To the novice, however, the above breakdown should provide some useful insight beyond the mere knowledge that a computer is a collection of electronic components. We note that different computers have different logical organizations, so that the scheme we have discussed embodies general ideas rather than the particular circuitry of any actual machine. A more detailed, but still accessible, exposition of how the hardware works in a serial computer is given in Section 9.1 of Arbib (1984).

Acknowledgment. Figures 2.3, 2.4, 2.6, 2.7, and 2.8 are adapted from Arbib (1973), with kind permission of Academic Press, Inc.

References for Chapter 2

Arbib, M.A., 1973, Automata theory in the context of theoretical neurophysiology, in *Foundations of Mathematical Biology* (R. Rosen, Ed.), pp. 191–282.
Arbib, M.A., 1984, *Computers and the Cybernetic Society*, Second Edition, Academic Press.
McCulloch, W.S., and Pitts, W.H., 1943, A logical calculus of the ideas immanent in nervous activity, *Bull. Math. Biophy.* **5**: 115–133.
Minsky, M.L., 1967, *Computation: Finite and Infinite Machines*, Prentice-Hall.
Shepherd, G., 1979, *The Synaptic Organization of the Brain*, Second Edition, Oxford University Press.

CHAPTER 3
Feedback and Realization

3.1 The Cybernetics of Feedback

The word "cybernetics" was coined by Norbert Wiener, 1948, and his colleagues to denote "the (comparative) study of control and communication in the animal and the machine." In a sense, then, all of the present book can be subsumed under the heading of "cybernetics." However, in this section we will be primarily interested in feedback and related issues in control theory, which form one of the central themes discussed by Wiener in his book.

In Section 2.1, we considered the nervous system as a three-stage system, as shown in Figure 2.1. To this end we now turn our attention to the arrow from effectors to receptors whereby information can be *fed back* to the nervous net proper as to how effectively it is controlling the activity of the effectors.

Let us make this clear by two simple examples.

1. If I wish to pick up a pencil, I move my hand toward it. My eyes then tell me how far my hand has to move. My brain thus continually gets information from my receptors (eyes) to tell me the position of my effector (hand). It can then compute appropriate instructions to issue to my arm muscles to move my hand in such a way as to reduce the difference between the actual and desired position of my hand.

2. When I walk, I alternately lift a foot and then let it fall. I control the muscle flexions in my leg to "change step" when the pressure exerted on my foot by the ground exceeds a certain critical value. Thus the position of my effector (foot) relative to the ground is *fed back* to my brain by my receptors (pressure-sensitive nerves in my foot). Once again, it is a difference—between the position of my foot and the ground—that is a

crucial determinant of the instructions that my brain issues to my leg muscles. The importance of this "feedback" from the pressure receptors of the feet is well illustrated by the following common experience: After sitting cross-legged for a long time, one gets up to find that one has "pins and needles" and has to walk without the use of tactile feedback. As a result, the leg movements are clumsy and uncoordinated.

The conclusion we draw from these two examples is that the arrow from effectors to receptors plays a crucial role in determining our responses. In other words, the concept of "feedback" must play an essential role in our study of brains and machines, where we say that an organism or a machine has feedback if its activity is controlled to some extent by the comparison of its actual performance with some tested performance. In particular, as our example of the hand and pencil showed, we are particularly interested in *negative* feedback, in which the machine uses the feedback to *decrease* the difference between actual and desired performance.

We may diagram a simple negative feedback system as shown in Figure 3.1(a). We continually feed into the system a quantity θ_d, which indicates the measure of the desired output. Now consider the actual output. We feed back its measure θ_0 to the error detector, which takes the difference between the desired and actual performances, and thereby calculates the error signal

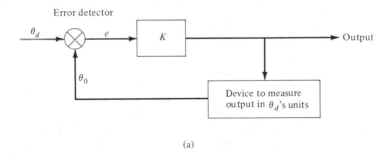

(a)

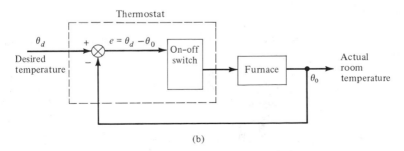

(b)

Figure 3.1   (a) A feedback system; (b) a thermostat.

1
$$e = \theta_d - \theta_0.$$

It is this error signal that actually controls the system K and so determines its output. The problem is to design the "black box" K in such a way as to make e go to zero. A simple example of negative feedback in machines is afforded by the thermostat, which serves to decrease the difference between actual performance (the temperature of the room) and desired performance (the temperature setting on the control of the thermostat) by suitably controlling the heat production of a furnace. The reader may easily verify that our thermostat may be put into the form of Figure 3.1(a), and is shown in Figure 3.1(b).

A negative feedback system of the type given in Fig. 3.1 is often called a *servomechanism*. I want to consider now a simple mathematical model of a *linear* servomechanism, where the relation between θ_0 and e is described by a linear differential equation (with constant coefficients). I give this example only to indicate how mathematics is used to describe feedback systems [a fuller account is given in Arbib (1988, Section 2.2)]—the only results of the model I shall use will be qualitative, and they will be "proved" intuitively later on. Consider now a very simple second-order *mathematical* model of K, as described by the equation (quickly derived in any introductory book on servomechanisms):

$$\frac{d^2 e}{dt^2} + 2\zeta\omega_n \frac{de}{dt} + \omega_n^2 e = \frac{d^2 \theta_d}{dt^2} + 2\zeta\omega_n \frac{d\theta_d}{dt},$$

where ζ and ω_n are constants characteristic of K.

Let us solve for the error $e(t)$ when our desired performance is a constant velocity output corresponding to $\theta_d(t) = k$. We obtain (see any standard text for the details)

$$e(t) = A + Be^{a_1 t} + Ce^{a_2 t},$$

where A, B, C are constants determined by ζ and ω_n and the various signal values in K at $t = 0$ (i.e., initially); where the e on the right-hand side is $2.17182818\ldots$; and where a_1 and a_2 are the roots of

$$a^2 + 2\zeta\omega_n a + \omega_n^2 = 0.$$

I have considered this second-order linear system because I want to show a qualitative property of servomechanisms—that of stability. For physical reasons, ζ and ω_n are real numbers, and so a_1 and a_2 are either both real or else conjugate complex numbers. If a_1 and a_2 are both real and negative, then $e^{a_1 t}$ and $e^{a_2 t}$ tend to 0 as t tends to ∞. However, if a_1, say, is positive, then $e^{a_1 t} \to \infty$ as $t \to \infty$, and in this case the system response is unstable. If a_1 and a_2 are conjugate complex numbers $a + ib$ and $a - ib$, then $Be^{a_1 t} + Ce^{a_2 t}$ can be put into the form

$$A + e^{at}(D\cos bt + E\sin bt),$$

which tends to A as $t \to \infty$ if a is negative, but yields unbounded oscillations

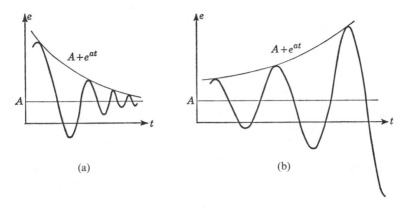

Figure 3.2   Error curves. (a) Stable: *a* negative; (b) unstable: *a* positive.

as $t \to \infty$ if *a* is positive (Figure 3.2). Thus, if either a_1 or a_2 has a positive real part, then the error is *unbounded* and the system is *unstable*. But if both a_1 and a_2 have negative real parts, the error is *bounded* and the system is *stable*.

This result—that a system which is not stable may enter unbounded oscillations—can also be understood in qualitative terms. Consider again Figure 3.1, where the purpose of *K* is to compensate for the error. Now suppose the malfunction of *K* consists in *overcompensation*. If the error *e* is initially positive, *K* overcompensates, and *e* becomes negative; *K* again overcompensates and *e* becomes large but positive; and so ad infinitum as the system goes into wild oscillations. Our desideratum in constructing a system, of course, is to so design *K* that the resultant servomechanism is stable, so that the error soon settles down to its final steady-state value *A* (hopefully zero).

The insight we have just gained into feedback and oscillation gives us insight into the disease of the nervous system known as ataxia. We quote from Wiener's *Cybernetics.**

> A patient comes into a neurological clinic. He is not paralyzed, and he can move his legs when he receives the order. Nevertheless, he suffers under a severe disability. He walks with a peculiar uncertain gait, with eyes downcast on the ground and on his legs. He starts each step with a kick, throwing each leg in succession in front of him. If blindfolded, he cannot stand up, and totters to the ground. What is the matter with him?
>
> Another patient comes in. While he sits at rest in his chair, there seems to be nothing wrong with him. However, offer him a cigarette, and he will swing his hand past it in trying to pick it up. This will be followed by an equally futile swing in the other direction, and this by still a third swing back, until his motion becomes nothing but a futile and violent oscillation. Give him a glass

*Reprinted from *Cybernetics*, Second Edition, by Norbert Wiener, p. 95, by permission of The MIT Press, The Massachusetts Institute of Technology, Cambridge, MA. Copyright 1948 and 1961 by the Massachusetts Institute of Technology.

of water, and he will empty it in these swings before he is able to bring it to his mouth. What is the matter with him?

Both of these patients are suffering from one form or another of what is known as *ataxia*. Their muscles are strong and healthy enough, but they are unable to organize their actions. The first patient suffers from *tabes dorsalis*. The part of the spinal cord which ordinarily receives sensations has been damaged or destroyed by the late sequelae of syphilis. The incoming messages are blunted, if they have not totally disappeared. The receptors in the joints and tendons and muscles and the soles of his feet, which ordinarily convey to him the position and state of motion of his legs, send no messages which his central nervous system can pick up and transmit, and for information concerning his posture he is obliged to trust to his eyes and the balancing organs of his inner ear. In the jargon of the physiologist, he has lost an important part of his proprioceptive or kinesthetic sense.

The second patient has lost none of his proprioceptive sense. His injury is elsewhere, in the cerebellum, and he is suffering from what is known as a cerebellar tremor or purpose tremor. It seems likely that the cerebellum has some function of proportioning the muscular response to the proprioceptive input, and if this proportioning is disturbed, a tremor may be one of the results.

We thus see that for effective action on the outer world it is not only essential that we possess good effectors, but that the performance of these effectors be properly monitored back to the central nervous system, and that the readings of these monitors be properly combined with the other information coming in from the sense organs to produce a properly proportioned output to the effectors.

Thus, the first form of ataxia is caused by the loss of feedback, whereas the second seems to be caused by an unstable K. With this dramatic example before us, it is clear that the neurophysiologist will often be interested in the brain as an automaton which controls another system, namely, the skeleto-musculature of the body. The brain thus must be appropriately structured to *control* the body, bringing it to an appropriate state for interactions with the world. To perform this function, it must be able to *observe* the different states of the body, so that its control signals may be appropriate to the current configuration. For example, when an animal is tracking its prey, its brain must solve for its body what system theorists call the *regulation problem* in which we have some desired trajectory in state space that we desire the system to follow, and we seek appropriate inputs to the system that will bring it as close to the trajectory as possible. As we have seen, *feedback* enters naturally because information about the behavior of the system must be fed back to the controller for it to adequately compute inputs that will bring about the desired behavior in the controlled system. If possible, we try to bring the state in some finite time to lie on the trajectory. However, as we have also seen, the system may be such that in bringing the state at all quickly toward the trajectory we end up with an overshoot and we may well get unstable oscillations. In that case, the appropriate strategy will be to avoid overshoot by not trying to bring the state to actually lie on the trajectory, but rather trying to bring it closer and closer to the trajectory with increasing time.

Let us use $\phi(t_1, t_0, q, x)$ to denote the state to which the system will be sent at time t_1 if it is in state q at time t_0 and the input $x(t)$ is applied for each time t from t_0 to t_1. We may then phrase the regulator problem formally,

2 The Regulator Problem. Let $\phi^*(t)$ be some desired state trajectory of the system to be controlled, and let $q \neq \phi^*(s)$ be some perturbation of this trajectory at time s. Then the *regulator problem* is to find a control (that is, an input) function x_q such that $\phi(t, s, q, x_q) \to \phi^*(t)$ as $t \to \infty$ (where we here use $\to$ in the sense of "tends to"), either converging in finite time, or else being "stable" and converging closer and closer as time goes by.

It is clear that the control function should depend on the actual state q at time s, and that is why we have denoted the control we seek by the notation x_q.

Of course, many other control problems are treated in system theory. One question might be how to choose inputs so as to get the system to some desired state in the "best" way possible. The word "best" is crucial here, and the solution to the problem will depend on our criterion of goodness. In some applications the best solution might be that which achieves the desired state in the minimum time, whereas in other cases the best solution might be that which uses the minimum "fuel." When it comes to analyzing systems as complex as a human, the question of criterion becomes extremely difficult and "satisficing" (to use the term introduced by Herbert Simon for "reaching a *satisfactory* state") will probably be more important than "optimizing."

Since, as we saw in Section 2.2, everything relevant about the past behavior of the system is summed up in its present state, in choosing the input that is appropriate at any given time t, the only information that we should use, in addition to the desired motion $\phi^*(t)$, is the state at that time. In other words, when we are given a desired trajectory, we should then be able to determine what is the most appropriate input at any time as a function of the state. This leads to our definition of a control law.

3 Definition. A *control law* is a map $k: Q \to X$ that assigns to the state $q(t)$ at time t the value $x(t) = k(q(t))$ as the input at that time.

Of course, many parameters of the system to be controlled, and many parameters of the desired motion, will enter into determining what control law is suitable. If we were to implement such a control law, we should need a setup like that shown in Figure 3.3(a) in which a measurement of the present state is "fed back" to the controller so that it may determine the appropriate input $x(t) = k(q(t))$.

However, such a setup is too much of an idealization. In general, the whole state vector is not available for outside measurement, rather the output vector is available, and it will usually contain at best partial information about the present state. For instance, in a normal dynamical system of the kind studied in Newtonian mechanics, we would expect the output of the system to

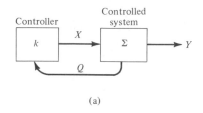

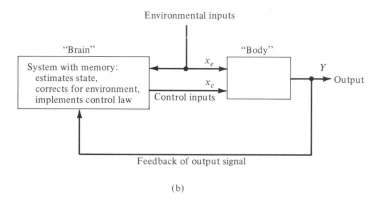

(b)

Figure 3.3   (a) Feedback of the entire state. (b) A more general situation in which the state must be estimated from more restricted feedback information.

consist of position measurements of its parts, whereas the full state vector must contain not only this position information, but also momentum or velocity information. We know that if we have the position information over any finite time period, then we may estimate the instantaneous velocity with arbitrary accuracy. Thus, in general, we must expect the controller—and the brain is no exception—to have memory and to use this memory to make an estimate of the present state of the controlled system (and, as our next point suggests, of the environment), given that it only has direct access to the system's outputs. One other consideration, which is certainly pertinent for an organism in a changing environment, is that it may not be possible for all the inputs to be under the control of the controller. When this is the case, one should have the controller monitor the environmental influences upon the system and then choose its present command on the basis of these environmental influences, as well as on the basis of its estimate of the present state of the system [see Figure 3.3(b)].

    If we are to have an efficent controller for a system, we must be assured that the system is *observable* in the sense that we can infer its state. As the example of estimation of velocity from a series of position measurements shows, while the controller may not be able to obtain the present state of the controlled system from its present output alone, it may well be able to obtain it if it can use a memory of a sufficient number of past observations of the output.

However, in other systems no number of output observations may suffice to enable us to estimate the present state. If we do not know the state, then we cannot hope to control it. (One point here: There may be some control problems in which we do not care to control all components of the state, and we may in fact be able to estimate all those components we wish to control. Then some more restricted notion of observability may be appropriate. However, in general, we will only include in the model those state variables that can affect those aspects of its behavior which will matter to us.)

Let us now say that two states are *indistinguishable* just in case we cannot tell from the way the system responds to its inputs which of the two states it was in.

4 Definition. Two states q and $\hat{q}$ are *indistinguishable* if we have that the outputs in response to the two states are equal for all input sequences x, for all times $t \geq \tau$,

$$\beta(\phi(t, \tau, q, x)) = \beta(\phi(t, \tau, \hat{q}, x)).$$

(Recall that β is the output map of the system.)

Thus, since we know that the design of a controller to solve an arbitrary regulator problem requires that we can estimate any state arbitrarily well, a necessary condition for the solution of the regulator problem will be that the system be *observable* in the following precise sense:

5 Definition. A system is *observable* if every pair of states is distinguishable.

Not only do we need *observability*. If we wish to get a system to go somewhere, we need to know where it is starting from; but we also require that the system be *reachable* in the sense that by applying suitable inputs every state of the system can be reached. (In particular control problems we shall not need reachability of every state, but perhaps only reachability of some particularly "desirable" subspace of the state space.)

6 Definition. A state q is *reachable* (from a state q_0) if there exist some times $s \leq \tau$ and some input function x such that

$$q = \phi(\tau, s, q_0, x).$$

There will be some control problems in which we do not even know the appropriate equations for describing the system we wish to control. Thus, the controller does not only have the task of estimating the state of a given system, but must also estimate what the actual equations of that system are. This problem of identifying the system under study is called *the identification problem*, and we shall study it in the next section. For the moment, let us simply note that in many *adaptive control* problems we may consider the identification problem the crucial one; we want the controller to change over time in such a way that, as a result of its interactions with the system to be controlled, it controls more effectively. In other words, it must adapt to the system it is controlling, and a crucial way of doing this is by identifying parameters that describe the system under control in a useful way.

Interestingly enough, we shall see in Section 3.2 that the identification problem of going from the input–output behavior of a system to its state equations is solvable if and only if the system is both reachable and observable. We shall also see, however, that given any system we may use its input–output behavior to identify a reachable and observable system with the same input–output behavior, our guarantee being that if the system under study is itself reachable and observable, then the system that we identify with identical input and output behavior will also have the same state behavior.

Let us go back to indistinguishability of two states. Suppose we are asked to transfer the system from a state q to an indistinguishable state $\hat{q}$. The answer is that we cannot if we are unable to tell which of the states we are in. But the reader might at this time comment, "Since you cannot tell which of the two states you are in, why should you ever care to transfer the system from one state to the other?" The reason for this is not a mathematical one, but a practical one. We must remember that the mathematical system we are looking at is an abstraction from a real system, and we may only expect this abstraction to accurately model the behavior of the system over a certain range. Thus, while we may expect two states to be indistinguishable as long as the model is applicable, they may have very different behaviors in that one state may lead to a behavior that takes us out of the domain of applicability, whereas the other does not. Clearly, then, we shall prefer to be in the latter state. For instance, if we have a system that contains a resistor which has no direct effect on the output of the system in normal functioning, then we might think that we could completely neglect the behavior of that resistor and thus ignore differences between two states when the only difference is in the value of the current passing through that resistor. That this is untrue is seen when we realize that if the resistor overheats, it will start to melt and foul up the system, moving us well out of the domain of applicability of our mathematical model. Similarly, a model of healthy brain function may fail to distinguish states that can yield crucial differences in neuroses. It is because of such considerations that we shall often wish to take careful note of differences between two states even when they are indistinguishable with respect to a given model of our system.

3.2 From External to Internal Descriptions

Even if we know completely the function, or behavior, of a device, we cannot deduce from this a unique structural description. However, it is comforting to know that mathematical theory can tell us, given the external description of the behavior of a system, what is the simplest internal structure that could yield the observed behavior. To go beyond this and find more about the actual system, if it is not the minimal system, we require something like anatomy; in other words, we need some clues about the type of internal structure before our theory can tell us the exact numerical details of the connections that must yield the output behavior.

Let us turn, in any case, to the *identification problem*: namely, the use of observations upon the inputs and outputs of a system to determine its internal structure. As we have already indicated, we cannot expect to determine the exact structure of the system under study. The process of going from the behavior of a system to its structural description is then not to be thought of as actually identifying the particular state variable form of *the* system under study, but rather that of identifying a state variable description of *a* system which will yield the observed behavior, even though the mechanism for generating that behavior may be different from that of the observed system.

Identification procedures are of real importance to the control theorist. For instance, suppose one wished to control a process, but did not know the dynamic equations of that process. Then, rather than build a controller specifically designed to control one actual system, we instead would build a general-purpose controller that could accommodate to any reasonable set of parameters to control the system given by that set of parameters, and inter-pose an identification procedure so that the controller would use at any time the set of parameters that the identification procedure provides as the best estimate of the real system's parameters (see Figure 3.4).

We note, without elaboration here, that it might also be necessary to have the identification procedure actually generate some of the input to the con-trolled process, in other words, apply test signals to try out various hypotheses about the parameters of the controlled process. One would then have to design a strategy to trade off the loss of optimality we get by not having a very accurate estimate of the state parameters against the loss of optimality we get by having the controller relinquish control to the identification procedure from time to time.

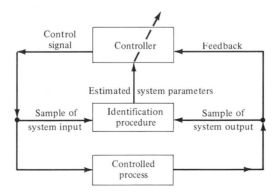

Figure 3.4 A controller made adaptive by the incorporation of an identification procedure.

In the above setup, we imagine that the controlled process has constant dynamics, so that as time goes by the identification procedure provides more and more accurate estimates of the system parameters to the controller, which can thus control the controlled process in a more closely optimal fashion. However, the above configuration may also prove useful when we have a controlled process that is slowly time variant; we might then imagine that the identification procedure could make accurate estimates of system parameters more quickly than they actually changed, thus allowing the controller to continue to act efficiently, despite the fluctuations in system dynamics. In other words, we couple an identification procedure to the controller not only to control a unknown system, but also to control a known system in which there is a drift in the parameters for which we must continually compensate. Note that the controller, when coupled to an identification procedure, is precisely what is often referred to as an "adaptive controller"; it adapts its control strategy to changing estimates of the dynamics of the controlled process.

If your task is to describe compactly what a system is doing, so that you can replace it with an equivalent system, then an identification procedure helps because it tells you the smallest system that will do the job, and you can then look at your equations and build this smaller consistent system. But if you are studying a real system—say, a biological system—the odds are, I presume, that it will not be minimal. Evolution has "decided" between alternatives, to some extent, but it cannot sit back and hone and polish everything individually to get the best possible system. However, once you have the minimal description, you can at least wire up a simulator so that when you carry out tests upon the system, you can see to what extent it does carry out some of the functions of the simulator, which is also a fairly compact way of expressing a model of the system against which you can try out other models; it condenses the data, and suggests in which "ball park" to look.

Let us now study how automata and system theory can solve the identification problem of going from the external behavior of a system to a description of its internal hebavior. We thus must make precise what we mean by an external description of a system, and see how it can be related to our notion of an internal description as given in terms of a next-state function in Section 2.2. But first we introduce the basic symbolism for describing sequences of input symbols.

Given any set X, we use X^* (read: X star) to denote the set of all finite sequences of elements from the set X. A sequence or *string of length n*,

$$x_1 x_2 \cdots x_n, \qquad \text{each } x_j \in X, \qquad 1 \le j \le n,$$

may be regarded as the result of n measurements of some variable that takes values in X. We include in X^* the "empty string" Λ of length 0; we may regard Λ as our "log sheet" before we have made any measurements.

Given two strings

$$w = a_1 \cdots a_m, \qquad \hat{w} = b_1 \cdots b_n,$$

we may form their *concatenation*

$$w \cdot \hat{w} = a_1 \cdots a_m b_1 \cdots b_n$$

comprising the measurements in w followed by those in $\hat{w}$. We thus have the equalities

$$w \cdot \Lambda = \Lambda \cdot w = w \quad \text{for all } w \in X^*$$

and

$$|w \cdot \hat{w}| = |w| + |\hat{w}| \quad \text{for all } w, \hat{w} \in X^*,$$

where we use $|w|$ to denote the *length* of w.

We may then define a *causal time-invariant system with input set X and output Y* in two distinct ways. The first we recognize as the notion of automaton from Section 2.2.

1 (*Internal Description*). By specifying a new set Q, and two functions $\delta: Q \times X \to Q: (q, x) \mapsto \delta(q, x)$ and $\beta: Q \to Y: q \mapsto \beta(q)$. The interpretation of these symbols is that the system is in some state from the set Q at successive moments of the time scale, and, furthermore, if the system is in state $q \in Q$ and receives input $x \in X$ at time t, then the output at that time will be $\beta(q)$, and the state at time $t + 1$ will be $\delta(q, x)$.

2 (*External Description*). By specifying a function $f: X^* \to Y$ that maps the set X^* of sequences of inputs to single outputs in Y. The interpretation is that if the system is activated in some standard state, and fed the sequence $x_1 \cdots x_n$ of inputs, then the output immediately subsequent to the processing of this string will be $f(x_1 \cdots x_n)$.

Linear system theory studies the case in which Q, X, and Y are all vector spaces and the maps δ and β are linear:

$$\delta(q, x) = Fq + Gx, \qquad \beta(q) = Hq,$$

for suitable linear transformations (matrices) F, G, and H. Automata theory usually considers Q, X, and Y to be finite sets, but allows δ and β to be nonlinear.

The passage from the internal description to the external description is immediate. First we extend δ to process any input string, $\delta^*: Q \times X^* \to Q$ by requiring

$$\delta^*(q, \Lambda) = q, \qquad \text{where } \Lambda \text{ is the empty string}$$

and then using $\delta^*(q, wx) = \delta(\delta^*(q, w), x)$ for all input strings w and inputs x, to recursively specify $\delta^*(q, w)$ for all input strings w.

Then let q_0 be the standard state in which the system is to be activated, and the external description f is immediately generated by the prescription

$$f(w) = \beta(\delta^*(q_0, w)) \quad \text{for all } w \text{ in } X^*.$$

Note, of course, that we may associate an external description M_q with *any* state q of a system M:

$$M_q(w) = \beta(\delta^*(q, w)) \quad \text{for all } w \text{ in } X^*.$$

We call M_q the *response function* of M started in state q.

The *identification problem*, then, is to effect the more complex passage from a function $f: X^* \to Y$ to an internal description of a *realization* of f, that is, a system M with a state q such that $f = M_q$.

To specify the sense in which the realization we seek is minimal, we first rephrase the definitions of two basic concepts of control theory, *reachability* and *observability*, given in Section 3.1:

Given a system M in "state-variable form," that is, with internal description (X, Y, Q, δ, β), and a specified initial state $q_0 \in Q$, we say that

(a) M is *reachable* if every state can be reached from q_0, that is, every $q \in Q$ can be written $\delta^*(q_0, w)$ for at least one $w \in X^*$; and that

(b) M is *observable* just in case any two distinct states yield observably different responses, that is, just in case for each pair $q \neq q'$ of distinct states in Q, there is at least one input sequence w to which they react differently: $M_q(w) \neq M_{q'}(w)$, so that the functions M_q and $M_{q'}$ are unequal.

In algebraic terminology, then, we say that M is *reachable* just in case the map $\delta^*(q_0, \cdot): X^* \to Q$ is *surjective*; and that M is *observable* just in case the map $\delta \mapsto M_q$ is *injective*. [See Kalman, Falb, and Arbib, (1969, Section 6.3) for a discussion of the fact that there are other definitions of observability, all equivalent for linear systems, which need not hold simultaneously for nonlinear systems.]

We shall now show that from f we can form a reachable observable system $M(f)$ with a state whose response function is f. Let us use w to denote some sequence in X^*, and $q \cdot w$ to denote the state to which the input sequence w sends state q of a system. To study the response function of this state $q \cdot w$, note that the sequence w sends us from the state q to the state $q \cdot w$, which w' then sends to $q \cdot ww'$. The output when we arrive at that state may be computed in two ways. We may consider it either as the response of a system started in state $q \cdot w$ and fed the sequence w' in which we denote it by $M_{q \cdot w}(w')$, or as the response of a system when started in state q and fed the complete sequence ww' in which case we denote it by $M_q(ww')$. Thus

$$M_{q \cdot w}(w') = M_q(ww')$$

for all states q of the system M, and for all input sequences w and w' in X^*.

Suppose that we have specified some initial state q_0 of our system M. A state is reachable from q_0 if and only if there exists some sequence w in X^* such that $q = q_0 \cdot w$. (This indicates why our identification problem will only yield a reachable system: If the only datum we have is some response function f such that $M_{q_0} = f$, then carrying out experiments upon this information can only yield data about states reachable from q_0. If M has states not reachable from q_0, our experiments can yield no information about them, since the only experiments we have at our disposal are studies of the external behavior of the system when we apply sequences to it, when started in state q_0.)

Now any two states reachable from q_0 must be of the form $q_0 \cdot w_1$ and $q_0 \cdot w_2$ for suitable input sequences w_1 and w_2 in X^*. These states are indistinguishable (see Section 1.3) if and only if they have exactly the same response functions, in other words, just in case $M_{q_0 \cdot w_1} = M_{q_0 \cdot w_2}$, that is, just in case

$$M_{q_0}(w_1 w') = M_{q_0}(w_2 w')$$

for all input sequences w' in X^*.

Thus, for *any* realization M of f, and *any* state q_0 of M for which $M_{q_0} = f$ holds, two input sequences w_1 and w_2 drive the system M from q_0 to indistinguishable states if and only if we have

$$f(w_1 w') = f(w_2 w')$$

for all input sequences w' in X^*.

Our strategy so far has been to assume nothing about the system M save that our experiments upon it start in some state q_0 whose response function $M_{q_0}: X^* \to Y$ satisfies the equality $M_{q_0} = f$ for some specified function f.

Let us assume for the moment that the system M with initial state q_0 such that $M_{q_0} = f$ is in fact both reachable and observable. Our analysis of the properties of such a system will, fortunately, show that such a system exists but, as yet, the existence of a reachable observable realization of the given f is still hypothetical. Reachability of M simply says that all states can be written in the form $q = q_0 \cdot w$ for at least one (and usually infinitely many) input sequence(s) w. Observability says that if two states are indistinguishable, then they must be identical. In other words, referring to our earlier discussion, we see that the condition

$$f(w_1 w') = f(w_2 w') \quad \text{for all input sequences } w' \text{ in } X^*,$$

which tells us that w_1 and w_2 lead us to indistinguishable states of *any* realization M of f, now tells us that w_1 and w_2 lead up to *identical* states of an *observable M*.

What we are saying, then, is that if M is an observable realization of f, then two strings lead us to the same state of M if and only if they are equivalent under the relationship E_f defined by

$$w_1 E_f w_2 \Leftrightarrow f(w_1 w') = f(w_2 w') \quad \text{for all input sequences } w' \text{ in } X^*.$$

Now note that E_f is an equivalence relation; by this we mean that E_f divides the set of input strings into disjoint classes and two input strings are in the same class if and only if they are equivalent. We write $[w_1]$ to denote the equivalence class containing the input string w_1, and then note that our definition of equivalence class simply says that

$$[w_1] = [w_2] \Leftrightarrow w_1 E_f w_2.$$

We use the notation X^*/E_f to denote the collection of all equivalence classes of input strings under the relationship E_f. In other words, we may write the equality

$$X^*/E_f = \{[w]|w \text{ in } X^*\},$$

the collection of equivalence classes of strings w in X^* under the equivalence relation E_f.

We should emphasize that there are many different labels for each element in the equivalence class, but an equivalence class only counts once. To make this more clear, let us consider the set $\mathbf{N}$ of all integers and the equivalence relationship $n_1 \equiv n_2$ which holds just in case $n_1 - n_2$ is an even number. It is then clear that there are only two equivalence classes, namely, the equivalence class of the even numbers and the equivalence class of the odd numbers, so that we may write $\mathbf{N}/ \equiv \; = \{\text{EVEN, ODD}\}$ and note such equalities as $[1] = [373]$, $[2] = [6]$ but also note the inequalities such as $[1] = \text{ODD} \neq \text{EVEN} = [2]$. Thus, when we write $\mathbf{N}/ \equiv \; = \{[n]|n \text{ in } \mathbf{N}\}$ we realize that although we have presented on the right-hand side infinitely many descriptions of the elements of $\mathbf{N}/ \equiv$, there are only two elements in that set; it is just that each element has infinitely many descriptions, ODD by all the odd numbers, and EVEN by all the even numbers.

So far, so good. We have seen that *if* there is a reachable realization of f, then its states can in fact be uniquely labeled by the elements of X^*/E_f; we identify the state $q_0 \cdot w_1$ with the element $[w_1]$. The question now is, "Is there such a system?" We shall show that there is by giving a mathematical description that actually has X^*/E_f for its state space.

Suppose that we are in the state $[w]$ and we apply x in X. What state will we then be in? Since $[w]$ is the state we can reach by starting in the initial state and applying input sequence w, and we move to the new state by applying input sequence x, it is clear that the new state should be $[wx]$, the state we reach by starting in the initial state and applying input sequence w followed by input x. For this definition of the next state to be valid, we must quickly check that the next state depends only on the equivalence class and not on the particular sequence we choose to represent it. In other words, we must show that if two input sequences w_1 and w_2 are such that $[w_1] = [w_2]$, then no matter what input x we shall consider we shall also have $[w_1 x] = [w_2 x]$.

Let us be quite clear as to why we must prove this. We know that if we do actually have any real system M and some state q_0 for which $M_{q_0} = f$, then it will of course be the case that if input sequences w_1 and w_2 take us to the same state $q_0 \cdot w_1 = q_0 \cdot w_2$, then it will certainly be the case that $w_1 x$ and $w_2 x$ will take us to the same state $q_0 \cdot w_1 x = q_0 \cdot w_2 x$. However, we have not taken an actual system; rather we have hypothesized that it *is* permissible to use X^*/E_f as the state space of a realization of f. In other words, we must explicitly verify that we may actually realize f by a system in which the states may be interchangeably labeled by sequences that belong to the same equivalence class with respect to E_f.

Let us now get on with the mathematical task of verifying that $[w_1] = [w_2]$

does indeed imply $[w_1 x] = [w_2 x]$ for any x in X,

$$[w_1] = [w_2] \Rightarrow w_1 E_f w_2$$
$$\Rightarrow f(w_1 w) = f(w_2 w) \quad \text{for all } w \text{ in } X^*.$$

But if the last equality holds for all sequences w in X^*, it certainly holds for all those sequences that start with our given input symbol x and so we may write

$$[w_1] = [w_2] \Rightarrow f(w_1 x w') = f(w_2 x w') \quad \text{for all } w' \text{ in } X^*,$$

but this latter condition is precisely that which ensures that

$$w_1 x E_f w_2 x, \quad \text{that is, } [w_1 x] = [w_2 x].$$

Thus, we may unambiguously define our next-state function on X^*/E_f by the rule $([w], x) \mapsto [wx]$.

It now remains to assign an output function. Suppose that we are in the state $[w]$. This is to say that it is a state that we could have got to from our initial state by applying input sequence w. But we specify that our initial state must be such that, whenever we plug in an input sequence w, the resultant output will be $f(w)$. In other words, this suggests that our appropriate output function is to be $[w] \mapsto f(w)$. Once again, it remains to verify that this does not depend on the choice of sequence that represents a state. But we have said that

$$[w_1] = [w_2] \Rightarrow w_1 E_f w_2$$
$$\Rightarrow f(w_1 w) = f(w_2 w) \quad \text{for all } w \text{ in } X^*,$$

and so this is certainly true in the case in which w is the empty string Λ that does not move us from the state to which we are sent by w_1 or w_2, respectively. In other words, if we take $w = \Lambda$, then $w_1 \Lambda = w_1$ and $w_2 \Lambda = w_2$, and so $[w_1] = [w_2]$ does imply $f(w_1) = f(w_2)$.

Finally, we should note that the initial state of our system is precisely that which we get to by not doing anything when we are in the initial state! In other words, our initial state is precisely that state $[\Lambda]$ that can be labeled by the empty string Λ. So, to complete our demonstration, let us verify that if we do start in that initial state, and put in an input sequence, then the state at which we arrive does indeed have the appropriate output:

(1) input x_1 sends state $[\Lambda]$ to state $[\Lambda x_1] = [x_1]$, which has output $f(x_1)$;
(2) input x_2 sends state $[x_1]$ to state $[x_1] \cdot x_2 = [x_1 x_2]$, which has output $f(x_1 x_2)$;
    $\dots$;
(n) input x_n sends state $[x_1 \cdots x_{n-1}]$ to state $[x_1 \cdots x_{n-1}] \cdot x_n = [x_1 \cdots x_n]$, which has output $f(x_1 \cdots x_n)$,

and so the output of our system after processing the whole sequence is indeed $f(x_1 \cdots x_n)$.

We thus have proved that we may always obtain an identification of a discrete-time response function by the procedure summarized in the following theorem [cf. Myhill (1957); Nerode (1958); and Rabin and Scott (1959)]:

Theorem. *Let f be any function $X^* \to Y$. In terms of f we may define a discrete-time system $M(f)$ by taking its state space to be*

$$Q_f = X^*/E_f,$$

where E_f is the equivalence relation on X^* defined by

$$w_1 E_f w_2 \Leftrightarrow f(w_1 w) = f(w_2 w) \quad \text{for all } w \text{ in } X^*;$$

its next-state function to be

$$\delta_f: Q_f \times X \to Q_f: ([w], x) \mapsto [wx];$$

and its output function to be

$$\beta_f: Q_f \to Y: [w] \mapsto f(w).$$

Then the system $M(f)$ is a reachable and observable realization of f, more specifically, f is the response function of the state labeled by Λ, the "empty string" of inputs:

$$f = M(f)_{[\Lambda]}.$$

The system is reachable because we only consider states that can be labeled by the input sequence that takes us to the given state from the initial state; and the system is observable because we decreed that whenever w_1 and w_2 led us to states indistinguishable by the experiments specified by function f, then they must lead us to the same state of $M(f)$. Note that if we take any system and carry out experiments upon it starting from some initial state, then the only observations that we can obtain will bear upon states reachable from the specified state.

We devote the remainder of the section to an informal discussion of the problem of structuring an appropriate internal state space on the basis of oversimplified descriptions of input–output behavior.

The brain, with its more than 100 billion neurons each capable of intricate patterns of electrochemical activity, is capable of a far more subtle dynamics than can be expressed in the usual ritualized, albeit extremely flexible, protocols of human language. The utterance of a word or movement of a limb is but a projection (in the mathematical sense of taking only a few coordinates of a complicated vector) of the state of the brain, and in analyzing the brain we must avoid the danger of trying to describe all of the brain's activity in terms that have evolved because of their appropriateness for describing output behavior. In any case, our descriptions also partake of being but one level of a hierarchy.

In conventional stochastic learning theory, one often studies learning tasks in a way which suggests that the brain has but two states, task learned or task unlearned, with nothing but random transitions to tie behavior together. Of course, the actual learning process in the real brain proceeds by numerous subtle changes; it is only the output that forces a binary value, masking the neural continuum. One is reminded of the situation in quantum mechanics. There we now describe the state of a system by a wavefunction that contains information about the probability distribution of results of measurement. But no measurement can tell us the wavefunction, and, even worse, the act of measurement may destroy information. In coming to describe the activity of the brain, we will have to evolve state descriptions as alien to present-day psychological jargon as the wavefunction is to the classical position and velocity description of Newtonian physics. Of course, Newtonian mechanics is perfectly adequate for a wide range of phenomena, and so may be much of conventional psychology. But as our powers of observation become more sophisticated, so must the inadequacies of the classical approach become more apparent.

A suggestive illustration comes from a computer simulation of the reticular formation (RF) of the brain stem. In mammals RF functions in complex interaction with higher brain centers and has as a primary responsibility for switching the organism from sleep to waking and back again. However, in lower animals, a case may be made for viewing the RF as a device for committing the whole organism to a mode of action. Certain anatomical considerations suggested a poker-chip analogy, in which local computations take place in thin sections normal to the direction of the spinal cord, and with connections between these sections or "modules" running up and down the RF. Each module receives a variegated sample of sensory information. The question is, "How can these modules interact in such a way as to reach a common decision to cause the animal to change into a suitable mode?" Kilmer et al. (1969) set forth a scheme of interconnections, which, when simulated on a digital computer, yield satisfactory switching characteristics. The point here is that the subsystems make tentative decisions on the basis of partial information, then slowly change under mutual interaction until sufficient consensus is reached to commit the organism to an overall mode of action. At no time would an "It's in this mode or that" description suffice to specify the dynamics of the model.

This discussion emphasizes that to understand complex behavior it will not be satisfactory to use the classical description of finite automata that specifies the next state for every state-input pair. Rather, we must see how various small automata are built up one upon the other in a hierarchical fashion, a topic that is, to some extent, discussed in the theory of schemas in *The Metaphorical Brain* (Second Edition).

Acknowledgment. Figures 3.1, 3.3, and 3.4 are adapted from Arbib (1973) (cited in Chapter 2), with kind permission of Academic Press, Inc.

References for Chapter 3

For a full textbook exposition of all the material in this chapter (with the exception of the brief asides on brain theory) see Padulo, L., and Arbib, M.A., 1974, *System Theory, A Unified State-Space Approach*, Saunders/Hemisphere Books. The items cited in Chapter 3 are as follows:

Arbib, M.A., 1988, *The Metaphorical Brain*: *An Introduction to Schemas and Brain Theory*, Second Edition, Wiley Interscience.

Kalman, R.E., Falb, P.L., and Arbib, M.A., 1969, *Topics in Mathematical System Theory*, McGraw-Hill.

Kilmer, W.L., McCulloch, W.S., and Blum, J., 1969, A model of the vertebrate central command system, *Int. J. Man-Machine Studies* 1: 279–309.

Myhill, J., 1957, Finite automata and the representation of events, WADD Technical Note 60–165, Wright-Patterson AFB, Ohio.

Nerode, A., 1958 Linear automaton transformations, *Proc. Amer. Math. Soc.* 9: 541–544.

Rabin, M.O., and Scott, D.S., 1959, Finite automata and their decision problems, *IBM J. Res. Dev.* 3: 114–125.

Wiener, N., 1948, *Cybernetics*: *or Control and Communication in the Animal and the Machine*, The Technology Press and Wiley. (Second Edition, The MIT Press, 1961.)

CHAPTER 4
Pattern Recognition Networks

4.1 Universals and Feature Detectors

How do we perceive the identity of a friend whether we see him in three-quarters face or in full face? How do we recognize a square whether it is large or small, near or far? How do we recognize a circle, even though it is not properly oriented and is seen as an ellipse?

One important factor in the comparison of the form of different objects is the visual–muscular feedback system. Some of this feedback is of a purely homeostatic nature, such as pupil dilation, which keeps light intensity within narrow bounds. When peripheral vision picks up one object conspicuous by brilliancy or by color or, above all, by motion, there a reflex brings the image into the fovea, that portion of the retina yielding best form and color vision. Thus we tend to bring any object that attracts our attention into standard position and orientation so that the visual trace of it formed in our nervous system varies within as small a range as possible.

However, this system of visual–muscular feedbacks is not sufficient to completely account for our perception of universals like chairs, circles, etc.* We must still account for the way in which our brains enable us to recognize a letter A, say, despite its being subjected to many transformations such as rotations, translations on the retina, shifts in perspective, etc. Pitts and McCulloch, 1947, provided the classic discussion of possible neural mecha-

---

*The *Oxford English Dictionary* tells us that a "universal" is "what is predicated of all the individuals or species of a class or genus; an abstract or general concept regarded as having an absolute mental or nominal existence; a general term or notion."

nisms that could account for this perceptual capability of our brains. Their designs show some interesting, although admittedly superficial, similarities to real brain structures.

We are now going to indulge in a little mathematics.

1 Definition. A *group of transformations* is a collection G of transformations such that if A and B are in G, then so also is AB in G (where AB is the transformation obtained by first carrying out B and *then* the transformation A; note that, in general, $AB \neq BA$); and where:

(a) I, the identity transformation, is in G. (I is the trivial "transformation" that leaves everything fixed, e.g., a translation through zero distance.)
(b) If A is in G, then it has an inverse A^{-1} in G, where $AA^{-1} = A^{-1}A = I$. (A^{-1} is the transformation whose effect is precisely to "undo" that of A).

Note that for transformations we automatically have associativity:

$$(AB)C = A(BC)$$

$$= C \text{ followed by } B \text{ followed by } A.$$

Pitts and McCulloch point out that

> ... numerous nets, embodied in special nervous structures, serve to classify information according to useful common characters. In vision they detect the equivalence of apparitions related by similarity and congruence like those of a single physical thing seen from various places. In audition, they recognise timbre and chord, regardless of pitch. The equivalent apparitions in all cases share a common figure and define a group of transformations that takes the equivalents into one another but preserve the figures invariant. So, for example, the group of translations removes a square appearing at one place to other places; but the figure of a square it leaves invariant. These figures are the "geometric objects" of Cartan and Weyl, and the "Gestalten" of Wertheimer and Kohler.

> We seek general methods for designing nervous nets which recognise figures in such a way as to produce the same output for every input belonging to the figure.

Now the things we see or hear are mapped within the brain as patterns of neuron firing in a cortical manifold (i.e., body of cells in the cortex) M. The distribution of excitation in M is described by a function $\phi(x, t)$, where $\phi(x, t) = 1$ if there is a neuron at position x firing at time t, and $\phi(x, t) = 0$ otherwise. [Where there is no risk of confusion, we omit the time variable and write $\phi(x, t)$ simply as $\phi(x)$.] Let G be the group of transformations that carry the functions $\phi(x, t)$ describing apparitions into their equivalents of the same figure and suppose G has N members. Pitts and McCulloch consider the case where the transformations T of G can be generated by transformations $\tilde{T}$ in the underlying manifold M, so that

$$T\phi(x) = \phi(\tilde{T}x),$$

i.e., the pattern is transformed by altering the active set of x's according to the transformation $\tilde{T}$.

For example, if G is the group of translations, then

$$T\phi(X) = \phi(x + a_T),$$

where a_T is a constant vector depending only on T. If G is the group of dilations

$$T\phi(x) = \phi(a_T x),$$

where a_T is a positive real number, the magnification factor, depending only on T. All such transformations are linear, that is

$$T(a\phi(x) + b\psi(x)) = a\phi(\tilde{T}x) + b\psi(\tilde{T}x)$$
$$= aT\phi(x) + bT\psi(x).$$

The simplest way to construct invariants of a given distribution $\phi(x, t)$ of excitation is to average over the group G. So let f be an arbitrary functional (i.e., a function of a function) that assigns a number to each $\phi(x, t)$. We form every transform $T\phi$ of $\phi(x, t)$, evaluate $f(T\phi)$, and average the results over G to obtain

$$a = \frac{1}{N} \sum_{\substack{\text{all} \\ T \in G}} f(T\phi).$$

If we had started with $S\phi$, S in G, instead of ϕ, we would have

$$\frac{1}{N} \sum_{\substack{\text{all} \\ T \in G}} f(TS\phi) = \frac{1}{N} \sum_{\substack{\text{all} \\ R \in G \\ \text{such} \\ \text{that} \\ RS^{-1} \in G}} f(R\phi)$$

$$= a, \quad \text{since } G \text{ is a group.}$$

To characterize completely the universal corresponding to the apparition that produces the neuronal firing pattern $\phi(x, t)$, we need a whole collection of such numbers a for different functionals f (one functional corresponding to "circularity," one to "squareness," etc.). We can distinguish between the different functionals by indexing them with a subscript ξ to yield f_ξ and letting ξ range over a set Ξ. We thus obtain the various averages

$$f_\xi(\phi) = \frac{1}{N} \sum_{T \in G} f_\xi(T\phi).$$

We shall actually introduce a new manifold of neurons, also called Ξ, with one neuron for each ξ. Ξ may split into several dimensions, in which case we can specify ξ by its coordinates $(\xi_1, \dots, \xi_m)$. If the neuron system needs less than complete information in order to recognize shapes, the manifold Ξ may be much smaller than M, have fewer dimensions, and indeed reduce to isolated points. The time t and some of the x_j representing position in M may serve as coordinates in Ξ.

Suppose the dimensions of Ξ are all spatial. Then the simplest neural net to realize the above formal process is obtained in the following way: Let the original manifold M be duplicated on $N - 1$ sheets, a manifold M_T for each T of G, and connected to M or its sensory afferents in such a way that whatever produces the distribution $\phi(x)$ on M produces the transformed distribution $T\phi(x)$ on M_T. The value of $f_\xi(T\phi)$ is computed by a suitable net, separately for each value of ξ for each M_T, and the results from all the M_T's are added by convergence on the neuron at the point ξ of the mosaic Ξ.

But to proceed entirely in this way usually requires far, far too many associative neurons to be plausible. The manifolds M_T together possess the sum of the dimensions of M and the degrees of freedom of the group G. More important is the number of neurons and fibers necessary to compute the values of $f_\xi(T\phi)$, which depends, in principle, on the entire distribution $T\phi$ and therefore requires a separate computer for every ξ, for every T of G. This difficulty is most acute if f_ξ is computed in a structure separated from the M_T, since in that case all operations must be performed by relatively few long fibers.

We can improve matters considerably by the following device:

Let the manifolds M_T be connected as before, but raise their thresholds so that their specific afferents alone are no longer able to excite them. Now let us introduce additional fibers ramifying throughout each M_T, so that when active they remedy the deficiency in summation and permit M_T to display $T\phi(x)$ as before.

Let all the neurons with the same coordinate x on the N different M_T's send axons to the neuron at x on another sheet exactly like them, say Q, and suppose any one of them can excite this neuron. If the additional fibers introduced above are excited in a regular cycle so that every one of the sheets M_T in turn, and only one at a time, receives the increment of summation it requires for activity, then all of the transforms $T\phi$ of $\phi(x)$ will be displayed successively on Q (cf. Figure 4.1).

A single f_ξ computer for each ξ, taking its input from Q instead of from the M_T's, will now suffice to produce all the values of $f_\xi(T\phi)$ in turn as the "time scanning" presents all the $T\phi$'s Q in the course of a cycle. These values of $f_\xi(T\phi)$ may be accumulated through a cycle at the final Ξ-neuron ξ in any way.

This device illustrates a useful general principle, which we may call the *exchangeability of time and space*. This states that any dimension or degree of freedom of a manifold or group can be exchanged freely with as much delay in the operation as corresponds to the number of distinct places along that dimension.

Impulses enter slantwise along the specific afferents, marked by plusses, and ascend until they reach the level M_a in the columns of the receptive layer activated at the moment by the nonspecific afferents. These provide sum-

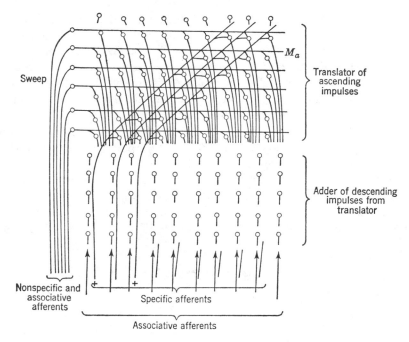

Figure 4.1 The Pitts–McCulloch scheme for translating input arrays and summing the result to produce group invariants as a possible scheme for extracting "universals."

mation adequate to permit the impulses to enter that level but no other (Pitts and McCulloch, 1947).

To the best of my knowledge, it has not been possible to prove neurophysiologically that the mechanisms presented above are actually employed in the human brain. However, the superficial resemblances are so great that Wiener was able to recount that when the neurophysiologist von Bonin saw a diagram like our Figure 4.1 he asked: "Is this a diagram of the fourth layer of the visual cortex?" What is perhaps more exciting is that *it was the search for such structures that led to the classic results on the frog visual system* described in the next section.

What the frog's eye tells the frog's brain

Inspired by the Pitts–McCulloch model of the visual brain as a layered computer, Lettvin, Maturana, McCulloch, and Pitts, 1959, addressed themselves to an investigation of the frog visual system designed to locate those properties of the system, if any, that assist the frog in recognizing the universals *prey* and *enemy*.

Frogs feed on insects, which they detect solely by vision. They prey only on moving insects, and their attention is never attracted by stationary objects. A large moving object provokes an escape reaction. For them, a form deprived of movement seems to be behaviorally meaningless. They seem to recognize their prey and select it for attack from among all other environmental objects because it exhibits a number of features such as movement, a certain size, some contrast, and perhaps also a certain color. This ability of frogs to recognize their prey and snap at it is not altered by changes in the general environment, e.g., changes in illumination.

Figure 4.2 shows the two eyes E of the frog, the optic nerves O proceeding back and crossing over, and the first portion of the brain they reach C, which is called the *superior colliculus* or *optic tectum*. (Other portions of the brain shown will not enter the present discussion.) The arrows give some idea of how the visual stimulus is mapped back to the colliculus.

Light enters the eye and passes to the back, where it enters the retina, passes through the transparent ganglion and internuncial cells, and finally reaches

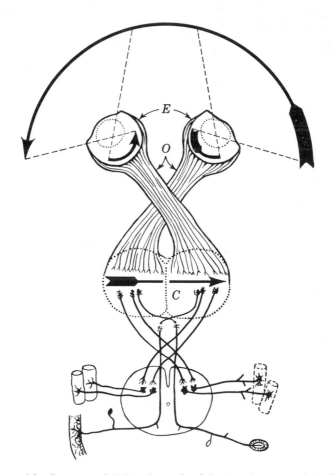

Figure 4.2   Ramon y Cajal's schematic of the visual system of the frog.

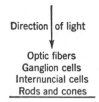

Figure 4.3   Passage of light through the frog retina.

the rods and cones (Figure 4.3). When sufficiently stimulated by light, the rods and cones produce generator potentials (continuously variable dc potentials), which, in turn, via the internuncial neurons, elicit impulses. The axons of these internuncial cells, in turn, impinge on the dendritic trees of the ganglion cells (neurons). The optic fibers are the axons of the ganglion cells—these fibers pass across the retina and come together at the blind spot, where they pass through the retina in a bundle called the optic nerve, which passes back to the brain.

The frog has about 1 million receptors (i.e., rods and cones), $2\frac{1}{2}$ to $3\frac{1}{2}$ million internuncial neurons, and $\frac{1}{2}$ million ganglion cells. Such numbers make it seem unlikely that the complex structure of the retina merely acts as a repeater that transmits intact to the brain the pattern of light and dark formed on the receptors. One is thus led to believe that the retina analyzes the visual image and transmits abstracted information to the visual centers.

Hartline, 1938, showed that the ganglion cells could be grouped into three classes, according to their response to a small spot of light on their receptive fields. (The *receptive field* of a ganglion cell is defined as the portion of the visual field mapped on the collection of all those receptors whose activity affects that of the given ganglion cell.) These classes were as follows:

(a) *On* cells, which respond with a prolonged but delayed discharge to the turning *on* of the spot of light;
(b) *On–off* cells, which respond with small bursts of high frequency to the change of *off* to *on*, or of *on* to *off* of that light; and
(c) *Off* cells, which respond with a prolonged and immediate discharge to the turning *off* of that light.

These observations were the first to show that the ganglion cells perform several complex operations on the visual image. But spots of light are not natural stimuli for the frog in the way that a fly or a worm is. Their use has suggested that the ganglions repeat to some extent, but in a coarser and inconstant manner, the original pattern of the visual image weighted by local differences. But the perception of universals that is obvious in the behavior of the frog seems to demand the presence of some functional invariants in the activity of the components of its visual system. That is, we expect that the retina might carry out computations on the visual image that serve to reveal

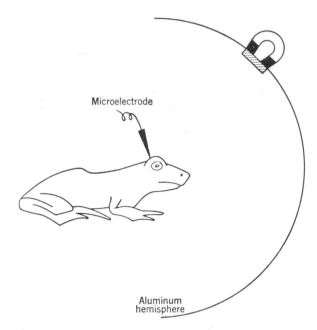

Microelectrode

Aluminum
hemisphere

Figure 4.4   Setup for study of the frog visual system.

crucial properties, such as the presence of a prey or an enemy. In fact, Barlow (1953) suggested that the on–off cells of Hartline might in fact be viewed as "bug detectors." Motivated by this finding, and by the theoretical structure of "layered feature detectors" suggested by Pitts and McCulloch (see Figure 4.1), Lettvin et al. sought to find appropriate functional invariants, and hence the functions of the ganglion cells, by adopting a naturalistic approach and studying the ganglion cells in terms of their response to real objects of the natural environment.

In their study they used the common American frog, *Rana pipiens*. The frog was placed so that one eye was in the center of a hemisphere 14 inches in diameter, which formed the experimental visual field (Figure 4.4). An electrode was inserted into the frog so that it could respond either to the activity of a single ganglion cell (by placing the tip of the electrode on the axon of the cell in the optic nerve) or to that of a tectum cell (Figure 4.2). The aluminum hemisphere represents about two-thirds of the visual field of one frog eye. By proper orientation of the animal, one could cover any desired part of the visual field and could entirely control the receptive field of the cells under study. The stimulating objects were moved on the inner surface by means of a magnet moved on the external surface. Numerous shapes and kinds of objects were used, e.g., dark disks, dark strips, and dark squares.

It was found that the response of ganglion cells fell into five groups:

Group I. The Boundary Detectors. These fibers have receptive fields 2–4° in diameter. They respond to any boundary between two shades of gray in the receptive field, provided the boundary is sharp. Sharpness of boundary rather than degree of contrast seems to be what is measured. The response is enhanced if the boundary is moved and is unchanged if the illumination of the particular contrast is altered over a very wide range. If no boundary exists in the field, no response can be got from change of lighting, no matter how sharp the change. Another property of the group I cells is that if a boundary is brought into the receptive field in total darkness and the light is switched on, a continuing response occurs after a short initial delay. Axons of group I are the "on" fibers of Hartline—a small, well-focused spot of light is defined by a sharp boundary.

Group II. The Movement-Gated, Dark Convex Boundary Detectors. These fibers have receptive fields of 3–5°. They too respond only to sharp boundaries between two grays, but only if that boundary is curved, the darker area being convex, and if the boundary is moved or has moved. Again, the responses are invariant over a wide range of illumination, roughly that between dim twilight and bright noon. They do not belong to any of Hartline's classes—they do not respond to spots of light, stimuli whose lighter areas are convex, which are not moved.

Group III. The Moving or Changing Contrast Detectors. These fibers have receptive fields 7–11° in diameter. They are, in effect, Hartline's "on–off" fibers. However, they respond invariantly under wide changes of illumination to the same silhouette moved at the same speed across the same background. They have no enduring response, but fire only if the contrast is changing or moving. The response is better (higher in frequency) when the boundary is sharp or moving fast than when it is blurred or moving slowly.

Group IV. The Dimming Detectors. These detectors are Hartline's "off" fibers. They have a 15° receptive field. They respond to any dimming in the whole receptive field weighted by distance from the center of that field. Boundaries play no role in the response. The same percentage of dimming produces the same reponse, more or less independent of the level of lighting at the beginning.

Group V. This group is rare. We cannot even say whether it has a receptive field in the usual sense. It fires at a frequency that is inversely related to the intensity of the average illumination. When the lighting is changed, the frequency slowly changes to its new level.

Each ganglion cell belongs to only one of these groups, and the cells of each class are uniformly distributed across the retina. In any small retinal area, one finds representatives of all groups in proportion to their general relative frequencies.

The axons of the cells of each group end in a separate layer of the tectum. However, two of them are mixed (the terminals of group V end in the strata of terminals of group III), so that they really form four fundamental layers of terminals. Each of these four layers of terminals in the tectum forms a "continuous" map of the retina with respect to the operation performed by the corresponding ganglion cells. The four layers are in registration, and at any point on a tectal lobe the terminals of all layers come from the same locus in the retina. We thus speak of each layer as forming a *retinotopic* (i.e., "place on retina") map.

Thus the function of the retina of the frog is not to transmit information about the point-to-point pattern of distribution of light and dark in the image formed on it. On the contrary, it is mainly to analyze this image at every point in terms of four qualitative contexts (boundaries, moving curvatures, changing contrasts, and local dimming) and a measure of illumination, sending this information to the tectum where these functions are separated in the four layers of terminals.

The retina transforms the visual image from a mosaic of luminous points to a system of overlapping qualitative contexts in which any point is described in terms of how it is related to what is around it. Since the transformation of the image constitutes the fundamental function of the retina, it is then the integrative capability of the ganglion cells that is significant. These considerations led Lettvin and his colleagues to an anatomical inquiry into the capacity of the ganglion cells to combine the information impinging upon them, seeking a correlation between the different morphological types of ganglion cells (types differing in the structure of their dendritic trees) and the operations that they perform. It is to their efforts in this inquiry that we now turn.

There are five anatomically distinct ganglion cells, as shown in Figure 4.5. You will note that the dendritic arrangements shown in the figure suggest that there are two major subdivisions of the inner plexiform layer (the layer of the

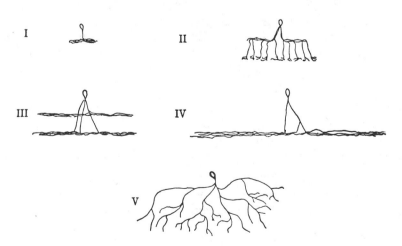

Figure 4.5   Five types of ganglion dendritic trees.

retina occupied by the dendrites of the ganglion cells), although undoubtedly each subdivision can be divided into several layers.

I. One-Level Constricted Field. These are the smallest of the ganglion cells. The major dendrites extend only to the inner levels of the inner plexiform layer and there spread out in dense and constricted planar bush.

II. Many-Level E Distribution. These are the next-to-smallest ganglion cells. The major dendrites extend only to the inner levels of the inner plexiform layer and there branch out in planar fashion. However, each branch emits twigs all along its course, and some extend into the outer levels of the inner plexiform layer, whereas others remain in the inner levels.

III. Many-Level H Distribution. These are the next-to-largest ganglion cells. The major dendrites extend to the outer levels of the inner plexiform layer. However, they emit two widely spread arbors, one in the inner levels and one in the outer levels.

IV. One-Level Broad Field. These are the largest of the ganglion cells. The major dendrites extend to the outer levels of the inner plexiform layer and there branch widely over a considerable area.

V. Diffuse trees. There are several sizes of these cells. The dendrites branch helter-skelter all over the inner plexiform layer and show no planar arrangement such as occurs in the other four kinds of cell.

If we should assume that the size of the dendritic field to some extent determines the size of the receptive field, we would emerge with a fairly definite correspondence between cell types and operations:

I. One-level constricted field	Boundary detection
II. Many-level E-shaped field	Movement-gated, dark convex boundary detectors
III. Many-level H-shaped field	Moving or changing contrast detectors
IV. One-level constricted field	Dimming detectors

More or less by default, we associate the diffuse type, which is rare, with the average light-level measuring group.

The above identifications are strengthened by these facts: The diameters of the dendritic fields match well the angular diameters of the receptive fields; the cell bodies are distibuted in size in the same way as the dendritic fields; and if the axon diameters reflect soma size, then the largest axons ought to have the largest receptive fields and the smallest axons the smallest fields, and this seems to be the case. Furthermore, receptive fields often appear to be not circular, but elliptical or cardioid. In Maturana's pictures of the ganglion cells, both elliptical and cardioid dendritic fields appear.

Turning from the ganglion cells to cells of the tectum, Lettvin and his colleagues found several kinds of cells. They were not able to define the subgroups at all well, but there are two major populations, which they named "newness neurons" and "sameness neurons." The former is concerned, it seems, with detection of novelty and visual events; the latter with continuity in time of interesting objects in the field of vision.

Comparisons

Embryology reveals that the retina is essentially a part of the brain. The work of Lettvin, Maturana, McCulloch, and Pitts, then, has laid bare some fundamental structure of the frog brain. It must, of course, be emphasized that their work applies only to the frog—neither the anatomy nor the receptive field operations are *necessarily* the same in other amphibia, and they are certainly not the same in mammals. Similar work on the cat's visual system has been done by Hubel and Wiesel (1962)—a basic paper which led to a thoroughgoing program of research on the visual system that was rewarded by a Nobel prize for Hubel and Wiesel. They found, in the cat's visual cortex, cells comparable to those that Lettvin et al. found in the colliculus of the frog. They comment:

> At first glance, it may seem astonishing that the complexity of third-order neurones in the frog's visual system should be equalled only by that of sixth-order neurones in the geniculo-cortical pathway of the cat. Yet this is less surprising if one notes the great anatomical differences in the two animals, especially the lack, in the frog, of any cortex or dorsal lateral geniculate body. There is undoubtedly a parallel difference in the use each animal makes of its visual system: the frog's visual apparatus is presumably specialised to recognise a limited number of stereotyped patterns or situations, compared with the high acuity and versatility found in the cat. Probably it is not so unreasonable to find that in the cat the specialization of cells for complex operations is postponed to a higher level, and that, when it does occur, it is carried out by a vast number of cells, and in great detail.

The comparison of preprocessing in frog and cat is carried out at greater length in Section 4.2 of *The Metaphorical Brain* (Second Edition); while Chapter 6 of that book is devoted to an exposition of models of Visuomotor Coordination in Frog and Toad, which carry forward the study of "What the Frog's Eye Tells the Frog's Brain" to investigate "What the Frog's Eye Tells the Frog."

4.2 The Perceptron

In Chapter 2, we glanced briefly at neurophysiology and abstracted our first model of the brain, dating back to 1943, the net of McCulloch–Pitts neurons. In the 1950s, Frank Rosenblatt and his colleagues at Cornell University developed a slightly different model of neural nets called the Perceptron (see

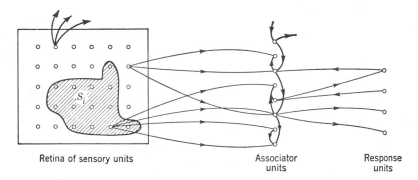

Figure 4.6   Schematic of a perceptron.

Rosenblatt, 1962, for a summary of this research). The main difference is that they did *not* make the assumption that the function of a neuron is fixed for all time. Instead, they allow the weights on each neuron to change with time. The purpose of this is to allow their neural net to change itself with time in such a way as to "learn" from experience. One may think of a perceptron as a pattern-recognition device that is not built to recognize a specified set of patterns, but rather has some ability to "learn" to recognize the patterns of a set after a finite number of trials.

The pattern is presented to the perceptron (Figure 4.6) on a *retina* of *sensory units* (e.g., photocells). Although the input need not be directly sensory, the motivation (as the terminology "retina" suggests) is to think of the pattern falling on the sensory units as a visual input of light and shadow. A photocell that receives a relatively light portion of the pattern is activated; one in a relatively dark portion is not. (This binary response to retinal illumination is, of course, simpler than physiological responses to nuances of color and illumination.) The sensory units are connected to *preprocessors*, also known as *associator units* (formal neurons), which in turn may be connected to each other or to *response units*.

In terms of our original neurophysiological consideration of the nervous system as a three-stage system, the retina constitutes the receptors of the perceptron; the associator units comprise the nervous system proper; while the response units correspond to the neurons whose output controls the effectors. It is again consonant with our original considerations that when a *stimulus* is presented to the retina of a perceptron, impulses are conducted from the activated sensory units to the associator units. If the total signal arriving at an associator unit exceeds its threshold, then the associator becomes *active* and sends impulses to the units to which it is connected.

In the setup shown in Figure 4.6, we view the preprocessor as a mechanism that extracts from the environmental input a set of d real numbers. The set will be called a pattern and the numbers components of the pattern. Any pattern can thus be represented by a point in a d-dimensional Euclidean space R^d

called the *pattern space*, where d is the number of measurements given by the preprocessor. The vector x of the measurements $(x_1, x_2, \ldots, x_d)$ can thus be used to represent the pattern. The pattern recognizer then takes the pattern and produces a response that may have one of N distinct values where there are N categories into which the patterns must be sorted.

In these terms a pattern recognizer is a function $f: R^d \to \{1, \ldots, N\}$. The points in R^d are thus grouped into at least N point sets, which we shall assume can be separated from each other by surfaces called *decision surfaces*. We shall assume for almost all points in R^d that a slight motion of the point does not change the category of the point. This is a valid assumption for most physical problems. The additional problem still exists of the category that is represented in more than one region of R^d. For example, a, A, *a*, and *A* are all members of the category of the first letter of the English alphabet, but they would probably be found in different regions of a pattern space. In such cases it may be necessary to establish a hierarchical system involving a computer apparatus that recognizes the subsets, and a separate system that recognizes that the subsets all belong to the same set. At any rate, let us avoid this problem by assuming that the decision space is divided into exactly N connected regions, eliminating split categories.

We call a function $g: R^d \to R$ a *discriminant function* if the equation $g(x) = 0$ gives the *decision surface* separating two regions of a pattern space. A basic problem of pattern recognition is the specification of such functions. (Sklansky and Wassel, 1981, give a textbook exposition of the state of the art.) Unfortunately, it is virtually impossible for humans to "read out" the function they use (not to mention *how* they use it) to classify patterns. What, for example, is your intuitive idea of the appropriate surface to discriminate A's from B's? Thus a common strategy in pattern recognition is to provide a classification machine with an adjustable function and to "train" it with a set of patterns of known classification that are typical of those that the machine must ultimately classify. The function may be linear, quadratic, or polynomial depending on the complexity and shape of the pattern space and the necessary discriminations. Actually the experimenter is choosing a class of functions with parameters which he hopes will, with proper adjustment, yield a function that will successfully classify any given pattern. For example, the experimenter may decide to use a linear function of the form

$$g(x) = w_1 x_1 + w_2 x_2 + w_3 x_3 + \cdots + w_d x_d + w_{d+1}$$

in a two-category pattern classifier. The equation $g(x) = 0$ gives a hyperplane as the decision surface, and training involves adjusting the coefficients $(w_1, w_2, \ldots, w_d, w_{d+1})$ so that the decision surface produces an acceptable separation of the two classes. We say that two categories are *linearly separable* if in fact an acceptable setting of such linear weights exists.

Thus far, then, a perceptron is another embodiment of grossly simplified neurophysiological data on a nervous system with purely visual receptors. However, the Perceptron group went further than this, and the additional properties of the net merit discussion here.

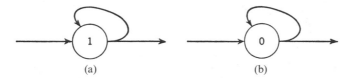

Figure 4.7   A hypothetical example of long-term memory.

Error-correction rule

There seems a great deal of evidence that humans have two kinds of memories —"short term" and "long term" It further appears that we have to retain an idea for quite a while in short-term memory before it is transferred into long-term memory. The time taken for this transfer has been variously estimated— one estimate is 20 minutes. It appears that if someone goes into coma, his memories of the 20 minutes or so prior to this are lost forever, i.e., they were not transferred to his long-term memory. It is now commonly believed that short-term memory is precisely that type of memory we gave our McCulloch–Pitts net—the passage of complicated patterns of electrical impulses through the net. It appears, then, that if such transient activity persists long enough it *actually changes the net.* This is best illustrated by a simple example.

The module of Figure 4.7(a) (cf. Figure 2.8) has a short-term memory of whether its input was ever fired, stored by the impulse reverberating in the loop. It would have long-term memory if the short-term memory could cause its threshold to drop from 1 to 0, for example—for the memory would then be preserved even if the reverberation should momentarily die down.

One postulated mechanism for long-term memory is formation of specific proteins within neurons, thus changing their thresholds in response to short-term memory patterns. Another mechanism is that endbulbs grow with repeated use, thus increasing the weight of the corresponding synaptic input and so, it might seem, making it easier to reestablish patterns of impulses using that synapse, hence making the corresponding memory easier to recall. But the precise mechanism is still unknown, and physiology, histology, and anatomy have yet to deliver a verdict. (But see Barnes, 1986a, 1986b, for a report on the encouraging progress made recently.)

The perceptron is equipped with a long-term memory. This is done by changing the weights of the inputs of neurons. These changes depend on the past activity of the termini of the connection. Rosenblatt called the rule of this dependence a *reinforcement* rule, but it is better called an *error-correction* rule, since it is designed to change the weights of the perceptron when it makes erroneous responses to stimuli that are presented to it. We refer to the judge of what is correct as the "teacher." The physiological evidence being so nebulous, Rosenblatt chose rules that were theoretically or experimentally convenient. The choices have enabled perceptrons to exhibit *aspects* of learning.

Rosenblatt had to decide how he would connect the sensory units to the associator units. A first approach was to make the connections *many-to-many*

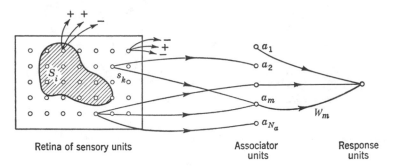

Figure 4.8   The simple perceptron with a single response unit. Connections from the retina to the preprocessing layer are fixed, but connections from the preprocessing layer to the linear threshold unit are variable via the adjustment scheme specified in the text.

and *random*. Later work introduced constraints into the network, e.g., subnets for line recognition like those found by Hubel and Wiesel in cat and monkey (and in the next section we shall see how *these* might be acquired). But just as our McCulloch–Pitts nets still had capacity for memory and computation despite the gross simplifications made in their derivation, so early perceptrons exhibited simple learning properties despite the assumptions made in construction. These properties form the material of the remainder of this section.

The reader may regard adaptive training as a case of the identification problem of Section 3.2. It is as if we were trying to find a model of a black box that classifies the patterns on the basis of some samples of its input–output behavior.

Consider the case of a twofold classification effected by using a threshold logic unit (i.e., McCulloch–Pitts neuron) to process the output of a set of binary feature detectors. This is an example of what Frank Rosenblatt called a *simple perceptron* with a single response unit (Figure 4.8). We then have a set R of input lines (to be thought of as arranged in a rectangular "retina" onto which patterns may be projected) for a network that consists of a single layer of preprocessors (associator units) whose outputs feed into a threshold logic unit with adjustable weights. A simple perceptron is one in which the associator units are not interconnected, *which means that it has no short-term memory.* If such connections are present, the perceptron is called *cross-coupled.* We want to analyze what classifications of input patterns can be realized by the firing or nonfiring of the output of such an array given different weight settings. The question asked by Rosenblatt and answered by many others since (an excellent review is in Nils Nilsson's, 1965, monograph on *Learning Machines*) is: "Given a network, can we 'train' it to recognize a given set of patterns by using feedback on whether or not the network classifies a pattern correctly to adjust the 'weights' on various interconnections?"

The answers have mostly been of the following type: "If a setting exists which will give you your desired classification, I guarantee that my scheme will eventually yield a satisfactory setting of the weights." Minsky and Papert, 1969, later revivified the study of perceptrons by responding to such convergence schemes with questions of the type: "Your scheme works when a weighting scheme exists, but *when* does there exist such a setting of the weights?" In other words, they ask, "Given a pattern-recognition problem, how much of the retina must each associator unit 'see' if the network is to do its job?" They analyze this question both for "order-limited perceptrons," in which the "how much" is the "number of input lines per component," and "diameter-limited perceptrons," in which the "how much" is the diameter of the input array from which each component receives its inputs. We shall say more about this in Section 4.4.

The Perceptron group has had three main modes of investigation: mathematical analysis (e.g., Block, 1962), simulation on a digital computer, and construction of an actual machine. Each method has its own advantages. One important result of using an actual machine is that it has been found that *neither precision nor reliability of the components is important, and the connections need not be precise.* Another interesting result is that the perceptron can "learn" *despite trainer error.*

The Mark I Perceptron, a hardware embodiment of a simple perceptron, had a retina of 20×20 photocells. Consider the case where we restrict our stimuli to the 26 letters of the alphabet, *each in standard position*—i.e., there are only 26 stimuli—and take our output from five binary-response units ($2^5 = 32 > 26$). In an actual experiment, the machine was reported to learn to identify them correctly after 15 exposures to each letter, a total of 390 exposures. But we require much more of a pattern-recognition system than that it discriminate between stimuli in standard position. What we want is a machine that can recognize each letter wherever it is placed on the retina, even though it be rotated slightly, distorted, and shown against a spotty background. But once we allow this, the number of stimuli jumps alarmingly, as may be seen by considering the motions of a standard letter E, presented as a pattern of dots as in Figure 4.9. To handle this, the simple perceptron is too

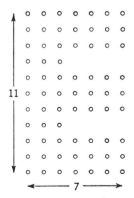

Figure 4.9   A letter E as it might be sensed by the sensory units of a perceptron.

simple (cf. Fukushima, 1980). We have discussed the simple perceptron here for pedagogical reasons—it illuminates a discussion of memory in neural nets. Before we leave this topic, it is only fair to warn the reader that the perceptron work received much bitter criticism. Certainly, early papers by the Perceptron group made exaggerated claims. However, one must be careful not to over-react, and we shall study a broader set of approaches in Chapter 5 which "vindicate" the Perceptron work, providing new algorithms which extend learning to multilayered networks. In fact, the present danger is again one of *over*enthusiasm.

The perceptron convergence theorem

We now provide a formal account of one of the perceptron error-correction schemes. First some notation: With each predicate ϕ, which asserts of a pattern x whether it is true or false that it possesses some property, we shall associate the binary function

$$\lceil \phi(x) \rceil = \begin{cases} 1 & \text{if } \phi(x) \text{ is true,} \\ 0 & \text{if } \phi(x) \text{ is false.} \end{cases}$$

For two vectors $\mathbf{x} = (x_1, \ldots, x_n)$ and $\mathbf{w} = (w_1, \ldots, w_n)$, we use $\mathbf{x} \cdot \mathbf{w}$ to represent the scalar product $(x_1 w_1 + \cdots + x_n w_n)$.

Suppose, then, that there are d feature detectors in the preprocessing layer in our simple perceptron (Figure 4.8), so that the input to the response unit is a vector $\mathbf{x} = (x_1, \ldots, x_d)$. Let us augment $\mathbf{x}$ by adding a $(d + 1)$st component equal to 1 to obtain $\mathbf{y} = (x_1, \ldots, x_d, 1)$. Then if we let $\mathbf{w} = (w_1, \ldots, w_d, -\theta)$, which is the weight vector augmented by minus the threshold θ, we see that the equation for the response r of the unit can be abbreviated from

$$r = 1 \quad \text{iff } \sum_{i=1}^{d} w_i x_i \geq \theta \quad \text{iff } \sum_{i=1}^{d} w_i x_i - 1 \cdot \theta \geq 0$$

to the simple form

$$r = \lceil \mathbf{w} \cdot \mathbf{y} \geq 0 \rceil.$$

Consider a finite set $\mathbf{Y}_1$ of augmented vectors corresponding to category 1, and a finite set $\mathbf{Y}_2$ of augmented vectors corresponding to category 2. We are going to specify a particular error-correction rule, and then show that *if* there is a setting of weights which will enable the response unit to discriminate the categories, *then*, with "sufficient experience" and using this learning rule, the response unit will achieve separation. In assuming that the two categories are linearly separable in this way, we guarantee that there exists at least one $(d + 1)$ weight vector $\hat{\mathbf{w}}$ such that

$$\mathbf{w} \cdot \mathbf{y} \geq 0 \quad \text{if } \mathbf{y} \in \mathbf{Y}_1,$$

$$\hat{\mathbf{w}} \cdot \mathbf{y} < 0 \quad \text{if } \mathbf{y} \in \mathbf{Y}_2.$$

We start with an arbitrary weight vector $\mathbf{w}$, which presumably misclassifies many patterns, and try to adjust it by repeated application of some error-correction training procedure. The procedure we shall study works repeatedly through the patterns in $\mathbf{Y}_1$ and $\mathbf{Y}_2$, testing each to see if the latest $\mathbf{w}$ classifies it correctly. Here, then, is our error-correction procedure: If $\mathbf{w}$ classifies the current $\mathbf{y}$ correctly, we leave $\mathbf{w}$ unchanged and move on to the next $\mathbf{y}$. If the classification is incorrect, however, we change $\mathbf{w}$ to $\mathbf{w}'$, where

$$\mathbf{w}' = \begin{cases} \mathbf{w} + \mathbf{y} & \text{if } \mathbf{y} \text{ belonged to category 1,} \\ \mathbf{w} - \mathbf{y} & \text{if } \mathbf{y} \text{ belonged to category 2.} \end{cases}$$

The idea is as follows: If $\mathbf{y}$ is in category 1 but $\mathbf{w}$ misclassified it, then we had $\mathbf{w} \cdot \mathbf{y} < 0$ where we should have had $\mathbf{w} \cdot \mathbf{y} \geq 0$. Since $\mathbf{y} \cdot \mathbf{y} < 0$ for any nonzero vector, we have that

$$(\mathbf{w} + \mathbf{y}) \cdot \mathbf{y} = \mathbf{w} \cdot \mathbf{y} + \mathbf{y} \cdot \mathbf{y} > \mathbf{w} \cdot \mathbf{y}$$

and so even if we do not have $\mathbf{w}' \cdot \mathbf{y} \geq 0$, we at least have that $\mathbf{w}'$ classifies $\mathbf{y}$ "more nearly correctly" than $\mathbf{w}$ does. Similarly for the category-2 correction.

Unfortunately, in classifying $\mathbf{y}$ "more correctly" we run the risk of classifying another pattern "less correctly." However, we shall now prove that Rosenblatt's procedure does *not* yield an endless seesaw, but will eventually converge to a correct set of weights *if one* exists, though perhaps after many iterations through the set of trial patterns.

To simplify the proof, we replace $\mathbf{Y}_2$ by $\mathbf{Y}_2' = \{-\mathbf{Y} | \mathbf{Y} \in \mathbf{Y}_2\}$. Then to say that $\mathbf{Y}_1$ and $\mathbf{Y}_2$ are linearly separable is to say that there exists a vector $\hat{\mathbf{w}}$ such that

$$\hat{\mathbf{w}} \cdot \mathbf{y} > 0 \quad \text{for all } \mathbf{y} \in \mathbf{Y} = \mathbf{Y}_1 \cup \mathbf{Y}_2'.$$

We may now rephrase our training procedure as based on $\mathbf{Y}$ rather than $\mathbf{Y}_1$ and $\mathbf{Y}_2$: Let $S_\mathbf{Y}$ be our *training sequence*, an infinite sequence of patterns such that it only contains patterns from $\mathbf{Y}$, and each of these occurs infinitely often. We then generate the sequence of weight vectors $\{\mathbf{w}^1, \mathbf{w}^2, \mathbf{w}^3, \ldots, \mathbf{w}^k, \ldots\}$ as follows, where $\mathbf{w}^1$ is arbitrary, and $\mathbf{y}^k$ is the kth pattern in the training sequence $S_\mathbf{Y}$:

$$\mathbf{w}^{k+1} = \begin{cases} \mathbf{w}^k & \text{if } \mathbf{w}^k \cdot \mathbf{y}^k > 0, \\ \mathbf{w}^k + \mathbf{y}^k & \text{if not.} \end{cases}$$

We want to prove that we eventually reach a weight vector $\mathbf{w}^k$ such that $\mathbf{w}^k \cdot \mathbf{y} > 0$ for all $\mathbf{y} \in \mathbf{Y}$, so that $\mathbf{w}^k = \mathbf{w}_0^k$ for all $k \geq k_0$.

Let $k_1, k_2, k_3, \ldots$ be the sequence of trials at which the weight vector is changed, and let us then denote $\mathbf{w}^{k_j}$ by $\hat{\mathbf{w}}^j$, and $\mathbf{y}^{k_j}$ by $\hat{\mathbf{y}}^j$. Then we have

$$\hat{\mathbf{w}}^j \cdot \hat{\mathbf{y}}^j \leq 0 \quad \text{and} \quad \hat{\mathbf{w}}^{j+1} = \hat{\mathbf{w}}^j + \hat{\mathbf{y}}^j$$

for all j, unless $\hat{\mathbf{w}}^j$ is already our sought-for terminal vector.

Taking $\hat{\mathbf{w}}^1 = 0$ (the reader may wish to modify the proof in the case of a nonzero initial weight vector), we then have

$$\hat{\mathbf{w}}^{j+1} = \hat{\mathbf{y}}^1 + \hat{\mathbf{y}}^2 + \cdots + \hat{\mathbf{y}}^j. \tag{1}$$

To prove that our correction procedure terminates, we must prove that j cannot be arbitrarily large in Eq. (1).

Let then $\mathbf{w}$ be any solution vector, that is, $\mathbf{y} \cdot \mathbf{w} > 0$ for all $\mathbf{y}$ in $\mathbf{Y}$. We can then define a positive number α by the equation

$$\alpha = \min\{\mathbf{y}' \cdot \mathbf{w} | \mathbf{y}' \in \mathbf{Y}\}. \tag{2}$$

Combining Eqs. (1) and (2) we deduce that

$$\hat{\mathbf{w}}^{j+1} \cdot \mathbf{w} = (\hat{\mathbf{y}}^1 + \cdots + \hat{\mathbf{y}}^j) \cdot \mathbf{w} \geq j\alpha.$$

Now the Cauchy–Schwarz inequality* says that for any 2 vectors $\mathbf{a}, \mathbf{b}$, we have $(\mathbf{a} \cdot \mathbf{b})^2 \leq |\mathbf{a}|^2 \cdot |\mathbf{b}|^2$ [where $|\mathbf{a}| = \sqrt{(\mathbf{a} \cdot \mathbf{a})}$ is the *length* of $\mathbf{a}$]. Thus, in the present case,

$$(\hat{\mathbf{w}}^{j+1} \cdot \mathbf{w})^2 \leq |\hat{\mathbf{w}}^{j+1}|^2 \cdot |\mathbf{w}|^2.$$

Therefore,

$$|\hat{\mathbf{w}}^{j+1}|^2 \geq \frac{j^2 \alpha^2}{|\mathbf{w}|^2}, \tag{3}$$

and so the squared length of the weight vector must grow at least quadratically with the number of steps.

We shall now show that such quadratic growth cannot continue indefinitely. Since $\hat{\mathbf{w}}^{j+1} = \hat{\mathbf{w}}^j + \mathbf{y}^j$ and $\hat{\mathbf{w}}^j \cdot \hat{\mathbf{y}}^j \leq 0$, we have for all j that

$$|\hat{\mathbf{w}}^{j+1}|^2 = |\hat{\mathbf{w}}^j|^2 + 2\hat{\mathbf{w}}^j \cdot \hat{\mathbf{y}}^j + |\hat{\mathbf{y}}^j|^2$$

$$\leq |\hat{\mathbf{w}}^j|^2 + |\hat{\mathbf{y}}^j|^2,$$

which yields, by repeated application,

$$|\hat{\mathbf{w}}^{j+1}|^2 \leq jM, \quad \text{where } M = \max\{|\mathbf{y}|^2 | \mathbf{y} \in \mathbf{Y}\},$$

which says that the squared length of the weight vector grows at most linearly with the number of steps. Thus, for each j we have

$$\frac{j^2 \alpha^2}{|\mathbf{w}|^2} < |\hat{\mathbf{w}}^{j+1}|^2 < jM$$

so that

$$j < \frac{M|\mathbf{w}|^2}{\alpha^2}.$$

Hence, the error-correction procedure must terminate after at most β steps,

---

*The proof is given in any text on linear algebra. For two-dimensional vectors $\mathbf{a} = (a_1, a_2)$, $\mathbf{b} = (b_1, b_2)$, this just says $(a_1 b_1 + a_2 b_2)^2 \leq (a_1^2 + a_2^2)(b_1^2 + b_2^2)$; it is equivalent to $2a_1 b_1 a_2 b_2 \leq a_1^2 b_2^2 + a_2^2 b_1^2$ which is just $0 \leq (a_1 b_2 - a_2 b_1)^2$, which is certainly true.

where β is the largest integer not exceeding $M|\mathbf{w}|^2/\alpha^2$, for any solution vector $\mathbf{w}$. Since every pattern occurs infinitely often in the sequence S_Y, termination can only occur if a solution vector is found, thus proving the theorem.

Therefore, we have the comfort of knowing that eventually our pattern recognizer will be able to classify correctly all the patterns in the training set. The next question of course is, "When is eventually?" The bound β is not very useful in estimating how many steps will be required in a given situation, since it depends on the knowledge of a solution vector $\mathbf{w}$, and the whole point of training is to find $\mathbf{w}$. It must also be remembered that the simplifying step of omitting from the training sequence those patterns that will not cause a change in the weight vector is impossible in an actual training situation. It is not even allowable to discard a pattern after it has once been correctly identified, for by the next time it comes up, the weight vector may have been moved so as to identify it incorrectly. Thus, the actual number of patterns that would have to be presented from the training sequence may greatly exceed β.

Implications

The anatomy and physiology of that portion of our brains involved in higher mental activity are little known. Although the gross anatomy of the brain reveals a complicated structure, the detailed anatomy yields a bewildering picture of seemingly random interconnections. It appears impossible that our genes specify the exact structure of our brains; rather, it is much more likely that they determine patterns of growth more or less open to the modifying effects of experience (cf. Section 4.3). Furthermore, even if the connections were strictly determined, we do not know the mechanism whereby the brain can recognize universals, e.g., recognize the letter A in many positions and despite many distortions. The reaction of the Perceptron group was to assume an initial randomness and allow all the structure requisite for pattern recognition to result from changes due to the error-correction rules. Their approach is interesting and yet, I feel, lacks something. There are intellectual acts open to a human child that are forever denied a gorilla—and these must be due, it would seem, to genetically determined differences in structure. Darwinian evolution took aeons to build the capability for pattern recognition into our brains—it would be surprising if a random network should evolve such a capability in a few hours of learning.

However, I must confess that all the above argument supports is the statement: "An adequate model of the human brain must be rich in a variety of specific structures, e.g., those involved in the perception of straight lines." The argument sheds no light on the question: "Is structure necessary for learning?" In other words, granted that the human brain possesses structure, we have still made no progress in resolving two conflicting points of view, namely,

1. Humans are intelligent because evolution has equipped them with a richly *structured* brain. This structure, while serving a variety of functions, in particular enables them to learn. A certain critical degree of structural complexity is required of a network before it can become self-modifying— no matter how sophisticated its reinforcement rules—in a way that we could consider intelligent.
2. Humans are intelligent because evolution has equipped them with a richly *interconnected* brain. The pattern of interconnections is irrelevant to truly intelligent learning, which results from the action of reinforcement rules on a sufficiently huge, but essentially random, net.

Perhaps the truth lies in a subtle blend of these views. The search for that blend is one of the most exciting quests in the study of the brain. We shall offer two formal approaches in what follows. In Section 4.4, we shall see how the structure of a network, and the complexity of the components, sets definite upper bounds on what can be learned by a network, *no matter what* the learning rule. Then in Chapter 5 we sample a number of learning rules that have been devised in the attempt to enable networks to learn as much as possible.

4.3 Learning without a Teacher

In the previous section, we studied Rosenblatt's Perceptron, in which a response unit accepts inputs from a network of preprocessors and adjusts its connections from them according to the following rule. Strengthen an active synapse if the response unit fails to fire when it should have fired; weaken an active synapse if the response unit fires when it should not. Thus, the learning rule needs a "teacher" to specify the correct response on each occasion. Here, we contrast *unsupervised Learning*, in which clustering of patterns that excite a cell is according to some built-in measure of similarity, without reference to any task-dependent measure of performance, with *supervised Learning*, in which the environment or "teacher" specifies the correct classification. (Associative nets use the desired recollection as the reference signal.) Discrepancy between what is done and what is required is used to update the weights.

Experiments on visual memory

Atrophy from disuse or misuse of visual mechanisms is well known in the clinic. Moreover, the experiments of Hubel and Wiesel, 1962, 1965, showed that monocular suturing of the eyelids or squint causes loss of binocularity of cells in the visual cortex of cats, whereas binocular suturing does not. But this leaves open the question of whether the effect is one of "disuse," or whether experience can actually create new "feature detectors." Hirsch and Spinelli

(1970) addressed this by presenting a kitten with limited visual exposure each day, in each case limited to patterns quite unlike single line segments (the Hubel–Wiesel stimuli)—a three-bar stimulus to one eye; a bullseye pattern to the other. They found that by limiting the visual experience of kittens in this way, to one or two simple visual patterns, it is possible to "fill up" their visual cortex with receptive fields whose shapes resemble the patterns in one or more details; some receptive fields are even recognizable, though blurred, representations of the image seen by the cat.

Shortly after the Spinelli–Hirsch experiment appeared in the literature, Blakemore and Cooper (1970) published a similar experiment, although with notable differences. They raised two kittens in the dark, except for a few hours every day when one kitten was put in a cylinder painted with vertical stripes and the other in one with horizontal stripes. Recording from the visual cortex after development showed that cells responded best to vertical lines, or lines close to vertical, in the vertically exposed kitten, but to horizontal lines, or close to them, in the horizontally exposed kittens. Units were binocularly activated. The almost photographic reproduction of images seen by the kittens in the receptive fields of neurons they studied in visual cortex suggested to Hirsch and Spinelli that they might be dealing with memory traces. Three possibilities had then to be examined:

1. The various classes of receptive fields one finds in the adult are genetically preprogrammed. Presence at birth, maturation after birth, or the necessity of environmental stimulations for the genome to express itself, are all subsumed in this hypothesis. The hypothesis can then be made that the unstimulated units atrophy, fail to mature, or are not expressed. This hypothesis demands clear and predictable behavioral deficits from the kittens described above. It also predicts that once the damage is done, it should be permanent, i.e., letting the kittens have further, normal experiences after the critical period should not change the physiological picture. Furthermore, the set of available classes can be reduced but not *changed*.

2. Cells are genetically preprogrammed as in (1); however, during the critical period partial or total reprogramming under environmental control can take place if needed. This would ensure that the animal has feature detectors optimal for the environment in which it finds itself, and that it at least has the set of detectors that has proved most helpful to its species through natural selection. The transience of the critical period would be an advantage since it would be impossible for the rest of the brain to interpret information from detectors whose coding properties change over time [cf. Kilmer and Olinski's (1974) core versus noncore scheme for training hippocampal circuitry]. This hypothesis has essentially the same predictions as the one above, i.e., clear behavioral deficits for tasks that demand nonexisting detectors and permanence of the physiological effects; however, it differs in that, even though there are limits, it allows for the property of generating receptive fields totally different from the ones originally preprogrammed.

3. This might be called a memory hypothesis, i.e., that there are no genetically programmed detectors: what is programmed is an adaptive network capable of storing, in a very direct fashion, elementary visual experiences. The receptive-field shapes in an adult cat would thus be bits and pieces of what the animal has seen in its past. This hypothesis predicts changes in the physiological picture, if new experiences are allowed, and also predicts no behavioral deficits, or perhaps only slight ones, since learning capacity does decrease with age.

In real life, of course, these three possibilities would be present in various ratios, depending on the animal. In fact, Fregnac and Imbert (1978) have shown that certain visual cortex cells are indeed "committed," others are slightly tunable, while yet a third does seems open to major restructuring by experience.

Models of visual memory

Spinelli's (1970) OCCAM (a computer model for an Omnium-gatherum Core Content-Addressable Memory in the central nervous system) suggests a scheme for how memories might be multiply stored in discrete neural networks. The basic neural circuit is reminiscent of spinal mechanisms: there a specific, prewired input pattern (e.g., an itch) elicits a specific, prewired behavior (e.g., a scratch). In OCCAM things are similarly organized, except that both the input and the output pattern are arbitrary and determined by experience. Furthermore, the model specifies how inputs "find their way" to the appropriate memory trace, i.e, the OCCAM networks are *content addressable*. Very little, if any, preprocessing of sensory activity is assumed. The model posits that the simplest and safest thing for an organism to do is to "store" information as it arrives, with the "selection" of "meaningful" stimuli accomplished by biasing memory at the time of action. Anything can thus become important and facilitate its own subset of memories.

Consider, then, the memory networks of Figure 4.10, addressed in parallel by stimuli entering the CNS. We shall see how they may form a content-addressable memory, wherein providing the system with part of a chunk of information will enable the system to play back the whole chunk.

The different columns are connected by collaterals from receiving cells and match cells, which carry lateral inhibition to other input cells in nearby networks. It is assumed that only one interneuron per column is active at any time and that different temporal segments of a pattern will be switched in a regular fashion through different interneurons. (In motor nerves, where individual fibers fire at about 10 per second, smooth contractions are obtained by regular phasing in and out of motor units.) A crucial assumption is that the synaptic conductivity tends in the limit to be directly proportional to the activity going through the synaptic junction itself, so that if a given quantity of activity is presented to the same synapse over and over again, an asymptote

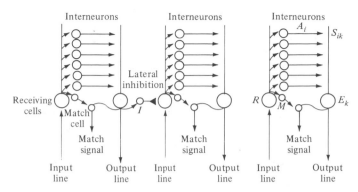

Figure 4.10   The basic wiring scheme of OCCAM. Each column learns a temporal sequence, with lateral inhibition blocking learning of a given sequence by too many columns.

will be reached such that the conductivity will represent faithfully the amount of activity that produced it.

We then assume that whenever a synaptic connection is activated, the amount of excitatory potential generated is proportional to the synaptic conductivity (not to the activity that has generated it). If a temporal pattern is presented to a network repeatedly, it will be stored in the synaptic conductivity of the interneurons, which will thus cause the output cell to play out a better simulacrum of the input pattern.

The match-cell output provides a measure of correlation in the activity of the input cell and the output cell from which it receives collaterals. The pattern is presented to all networks in parallel. But eventually one cell will have a somewhat better adjustment than its neighbors, the activity in its match cell will rise and (thanks to the lateral-inhibition mechanism) turn down the input to nearby networks—i.e., the network that gets ahead, by chance, draws the pattern to itself and prevents the other networks from learning it. The number of networks that learn the same pattern is thus determined by the extent of the lateral inhibition. To ensure that the lateral inhibition "gets there in time," it is assumed that each time a cell is activated, the more an afterdischarge, the more the synaptic conductivity will be changed. Thus the longer the after-discharge, the faster the learning—and this afterdischarge will be cut off by lateral inhibition if another "column" has already learned the pattern.

When one or more patterns have been stored, it is desirable that if new patterns are to be stored, this be done by networks that have not previously been used. To do this a given match cell becomes harder to activate if its network has been used often before. Then a pattern would have above-chance effect on a network in which it is already stored, reasonable effect upon an "uncommitted" network, and little effect on a network containing a well-learned pattern.

Pribram, Spinelli, and Kamback (1967) have suggested that presentation of a stimulus will generate a playback of the whole sequence: recognition of the stimulus; the appropriate behavior that goes with the stimulus; and, finally, expectation of the consequences of that behavior. The less stimulus presented, the more information there is in the playback, and the higher the risk there is in using it. (Of course, an animal might generate certain actions as an experiment to see if the present situation accords with the recalled details of the stimulus before reacting to the stimulus per se.) Ideally, then, the acceptable-match parameter should be set for the minimum value that allows unequivocal recognition of the stimulus, and thus the playback of the rest of that memory package containing information about what to do or not to do with it and what it expect.

Whereas the OCCAM model of 1970 has a temporal–spatial converter to enable it to record temporal patterns, the von der Malsburg model (1973) is closer to the spirit of the results on visual-pattern memory mentioned earlier in this section. It was formulated to determine whether a simple circuit, possessing only a few characteristics of the cat's visual system, would organize itself into the "simple-cell" receptive-field patterns found by Hubel and Wiesel in area 17 of cat visual cortex, each cell having one preferred orientation to which it responds maximally. Von der Malsburg thought that genetically specified patterns of lateral excitatory and inhibitory influences in the geniculostriatal system might highly predispose this system toward its columnar organization. He also thought that a loose genetic specification of the details of the retinogeniculostriatal projection could be coupled with plastic synaptic mechanisms in the projection so as to enhance the columnar organizational tendencies of the striatal system.

His model to test these ideas consisted of a retina of 19 binary element (A cells), a geniculostriatal manifold of 169 excitatory cells (E cells) and 169 inhibitory cells (I cells), interconnected according to a simple geometric specification. (The inhibitory interconnections serve much the same role as the lateral inhibition in OCCAM, with the E cells being both the "match cells" and the "output cells" since "recognition" rather than "temporal playback" is the task here.) The connection from each retinal cell A_i to each E cell E_k is through a Hebb synapse of strength S_{ik}, which is modified by $\Delta S_{ik} > 0$ for each "modification time step" if A_i is active *and* if E_k fires. In this case the magnitude of ΔS_{ik} is proportional to the firing rate of E_k.

A crucial problem of the Hebb rule is that all ΔS_{ik} are positive, with the consequent risk that all synapses will increase to a maximum, with cortical cells responding to "everything." von der Malsburg's solution was to renormalize his circuit to make the sum of synaptic weights equal to a specified constant after each "modification time step." A number of authors have offered solutions to this problem. [Milner (1957) was perhaps the first; Grossberg (1982) contains a collection of important papers on "neo-Hebbian" learning.]

Von der Malsburg used nine different retinal inputs, each corresponding to a "line" of different orientation. He presented them in an appropriately mixed

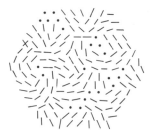

Figure 4.11   Each bar of this view onto the cortex indicates the optimal orientation of the E cell. Dots without a bar are cells that have never reacted to the standard set of stimuli. From von der Malsburg (1973).

order, and after 100 modification times froze the ΔS_{ik} and checked the retinal input orientations that elicited maximal responses in each E cell. Figure 4.11 shows the result. The clustering of units responding to a given orientation is strikingly reminiscent of Hubel and Wiesel's columnar mapping results. In later papers (e.g., 1979) von der Malsburg provided explicit analyses of the self-organization of this columnar organization.

Lateral inhibition, which was not mentioned in the basic experiments of Hubel and Wiesel, plays a crucial role in the models, so that different "experiences" are stored in different units. In such systems feature detectors arise only to the extent that they respond to commonly occurring aspects of the animal's experience. This well-known approach to learning has recently been given a clear presentation, with a historical introduction, by Rumelhart and Zipser, 1985. A pattern feeds cells connected by mutual inhibition, and eventually one cell only remains active as respondent to the pattern. [See Didday (1970) for an early example of such a "maximum selector" or "winner take all" network. Amari and Arbib (1977) give a mathematical analysis.] The "winner" adjusts its synapses by a Hebblike rule with normalization. They provide a formal treatment by conceptualizing the normalized weights and the input patterns as points on a sphere. When a pattern turns on, the learning rule moves the closest unit toward it on the sphere. The response of units will thus tend to define a cluster, with respect to a group of nearby patterns, through unsupervised learning.

Cooper, Liberman, and Oja (1979) have elaborated the von der Malsburg model, addressing a body of data, due to Imbert and his colleagues who, as mentioned above, classify cells in visual cortex as *specific* (specifically tuned to bars), *nonspecific*, and *immature* (with rectangular receptive fields, *weak* orientation selectivity, and better response to edges than to spots). Imbert et al. find the curves of Figure 4.12 for normal and dark-reared kittens. The suggestion is that there is some genetically programmed specificity that is lost unless augmented by environmental structure.

The model only incorporates an overall path from retina to cortex, with Hebbian (excitatory) synapses from retina to cortical cells and lateral inhibi-

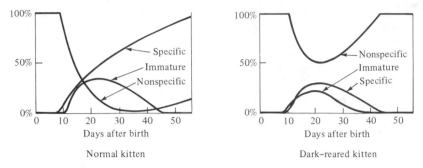

Figure 4.12   Curves showing the percentage of cells in kitten visual cortex that are specific, nonspecific, and immature as a function of days after birth in a normal (left) and dark-reared (right) kitten. (From Cooper et al., 1979.)

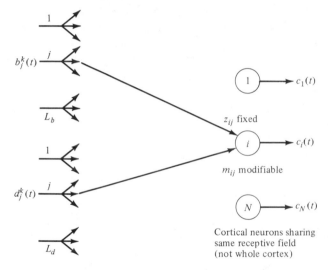

Figure 4.13   The connection scheme used by Cooper et al. in which the connection weights z_{ij} are fixed, while the weights m_{ij} are modifiable.

tion between cells of cortex. The Hebbian plasticity allows the embedding of commonly presented patterns; the lateral inhibition restricts the number of cells tuning to a given pattern. The input is segregated into two parts, one via fixed synapses z_{ij} and the other via plastic synapses m_{ij} (Figure 4.13). The $(b^k(t), d^k(t))$ vectors correspond to the kth "standard input vector" e^k—corresponding, e.g., to different bar orientations. Thus the b^k and d^k patterns, which are already transformed (e.g., in the thalamic "way-station" for vision, the lateral geniculate nucleus) need not themselves be barlike.

Now assume that cells are preloaded with b_k's that predispose a cell toward a pattern. Then without noise or lateral inhibition one obtains the curves of Figure 4.14.

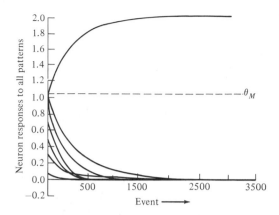

Figure 4.14 Curves showing that the system of Cooper et al. increases selectivity for output values above β and decreases it for values below β. (From Cooper et al., 1979.)

Like other workers, Cooper et al. must modify the Hebb scheme to stop all synapses saturating. Their ploy is to introduce two thresholds, a saturation limit μ on firing frequency and a modification threshold β. Then the learning role adjusts m_{ij} by the amount ∂m_{ij} where

$$\partial m_{ij} \sim \begin{cases} (\text{output}_i)(\text{input}_j) & \text{when the } \beta < \text{output} < \mu, \\ 0 & \text{when output} = \mu, \\ -(\text{output}_i)(\text{input}_j) & \text{when output} < \beta, \end{cases}$$

all smoothed out to a continuous curve.

Thus the system increases selectivity above β and reduces it below β. As in the other models we have discussed, this system uses linear feedforward lateral inhibition so that one highly responsive cell can help drive others below β. By preloading cells with b_k's which predispose a cell toward a pattern, these authors find that their learning rule yields the curves of Figure 4.14.

To conclude our discussion of "learning without a teacher," we briefly mention two further topics:

Kandel (e.g., 1978) studied the habituation of the gill withdrawal reflex in *Aplysia*—if the gill of the animal is touched, the animal withdraws its syphon reflexively; but with repeated stimulation, the animal ceases to respond. Kandel and co-workers have elegantly shown that this is due to a reduction in efficacy at certain synapses, and have begun to chart the biochemical mechanism underlying this change. There is a specific "parameter"—the specific synaptic pathway for the reflex—that is (reversibly, please note) "turned to the off position." This is, certainly, a very simple form of learning. However, Sokolov (1975) has shown that there are more subtle types of habituation which depend on learning a specific stimulus pattern, not on turning off a "switch." For example, a subject may habituate to a tone of a given frequency, but will show the orienting response to a tone of that frequency that is shorter in duration. The system builds a model of the stimulus, and compares the model with the current stimulus, reacting only if there is a mismatch [cf. Lara (1983) for a neural model of these processes].

Recent results in the study of neural development show that the genetic program for brain growth is open to experience; the theory of growth is itself beginning to develop to address the subtleties of neural development in a way that may begin to make contact with our understanding of *real* learning. We have already seen that Hirsch and Spinelli showed that cats could be so trained that visual cortex neurons would be specified for "new" features not present in the normal animal. More recently, Spinelli and Jensen (1979) have shown that the allocation of cells to different subsets of the sensory world can be modified on the basis of early experience. The theoretical models of this phenomenon by Amari (1980) and Kohonen (1982), make it clear that we have a situation in which innate structure provides the basis for, rather than precluding the operation of, powerful learning mechanisms.

4.4 Network Complexity

Evolutionary advantage may accrue to an organism from changes in the *metabolic* range of the organism with new enzymes allowing the organism to utilize new materials as foodstuffs; from gross *structural* changes such as limbs better adapted for motion in a certain type of environment; and from changes in *information-processing* capabilities of its nervous system. An animal can survive better in an environment if it not only has receptors that can detect an enemy far away but if it can compute on that information to find appropriate avoiding action before the enemy is upon it. Thus, we try to analyze how, for an animal with a given metabolic machinery and a given gross structure, changes in its receptor organization and its ability to combine present information with past experience can best contribute to its improved handling of its environment.

With this in mind, we can see that a crucial question for automata theory must be the study of complexity of computation to understand, for instance, how long a network of given components must take to compute a certain function or what the range of functions is that can be computed by networks of a given structure. We shall give some examples of such theory, but we should stress that the theory remains far removed from the full subtlety of the biological situation.

We shall place our study of computational complexity of networks in the context of pattern recognition. To survive, an organism needs to gain knowledge of the world around it. Such knowledge can never be of the actual object but can only be of a number of abstractions from the reality of the object, extracted from the flow of energy bathing the organism and modified by surrounding objects. The organism's survival depends greatly on its ability to classify sensory patterns as belonging to a category, and then basing its actions on the previous experience with that category (Section 4.2). The pattern recognition must be based on a limited number of measurements on the environment.

A common schematic in pattern recognition theory is to feed the values of a relatively effective set of parameters into a *preprocessor* (recall Figure 4.6). At this preprocessing stage most of the information coming in from the environment is lost, with only that information being transmitted that is necessary for classification. If the preprocessor deletes too much, recognition will be impossible or incorrect; if it does not delete enough, the recognizer may be swamped and unable to function.

With this, let us turn to two mathematical theories, relating network complexity to network function, realizing that at present it cannot be "plugged in" to solve biological problems, but may help us refine the questions we ask of the experimenter, and suggest important new ways of interpreting his results.

The Minsky–Papert theory

In our study of training a perceptron to make a binary classification, we noted that if a linear separation exists, the training algorithm will reach it. We now turn to Minsky and Papert's (1969) study of when it is possible for a McCulloch–Pitts neuron (no matter how trained) to combine the information in a single preprocessing layer to perform a given pattern-recognition task. We are going to be interested in pattern predicates ϕ such that $\phi(X)$ is true for some patterns X and false for others (e.g., "X is connected," i.e., a continuous path can be drawn from any one point of X to another, or "X is of odd parity," i.e., contains an odd number of active elements), and we shall ask such questions as, "How many inputs are required for the preprocessing units of a simple perceptron (Figure 4.8)?" Of course, we can always get away with using a single element, computing an arbitrary Boolean function, and connecting it to all the sensory units of the retina. So the question that really interests us is, "Can we get away with a small number of response units connected to each of the preprocessors?", to make a global decision by synthesizing an array of local views.

Minsky and Papert show that if a predicate is unchanged by various permutations of its inputs, then we may use this fact to simplify the weights on the response unit, and that this simplified form will often enable us to place a lower bound on the complexity of the preprocessors. Before giving the theory, we shall try to clarify the approach by a simple example.

Consider the simple Boolean operation of addition modulo 2 (Figure 4.15). If we imagine the square with vertices $(0, 0)$, $(0, 1)$, $(1, 1)$, and $(1, 0)$ in the Cartesian plane, with (x_1, x_2) being labeled by $x_1 \oplus x_2$, we have 0's at one diagonally opposite pair of vertices and 1's at the other diagonally opposite pair of vertices. It is clear that there is no way of interposing a straight line such

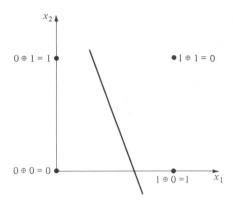

Figure 4.15   An intuitive demonstration that binary addition cannot be realized by a threshold element—no single line can separate the 0's from the 1's.

that the 1's lie on one side and the 0's lie on the other side. In other words, it is clear in this case—from visual inspection—that no threshold element exists that can do the job of addition modulo 2. However, we shall prove it mathematically in order to gain insight into a general technique used by Minsky and Papert.

Consider the claim that we wish to prove wrong—that there actually exists a neuron with threshold θ with weights α and β such that $x_1 \oplus x_2 = 1$ if and only if $\alpha x_1 + \beta x_2 \geq \theta$. The crucial point is to note that the function of addition modulo 2 is symmetric; therefore, we must also have $x_1 \oplus x_2 = 1$ if and only if $\beta x_1 + \alpha x_2$ exceeds θ, and, so, adding together the two terms, we have $x_1 \oplus x_2 = 1$ if and only if

$$\frac{\alpha + \beta}{2}x_1 + \frac{\alpha + \beta}{2}x_2 \geq \theta.$$

Writing $\frac{1}{2}(\alpha + \beta)$ as τ, we see that we have reduced three putative parameters α, β, and θ to a pair of parameters τ and θ such that $x_1 \oplus x_2 = 1$ if and only if $\tau(x_1 + x_2) \geq \theta$. We now set $t = x_1 + x_2$ and look at the polynomial $\tau t - \theta$. It is a degree-1 polynomial, but note: at $t = 0$, $\tau t - \theta$ must be less than zero $(0 \oplus 0 = 0)$; at $t = 1$, it is greater than or equal to zero $(0 \oplus 1 = 1)$; and at $t = 2$, it is again less than zero $(1 \oplus 1 = 0)$. This is a contradiction. A polynomial of degree 1 cannot change sign from positive to negative more than once. We thus conclude that there is no such polynomial, and thus that there is no threshold element which will add modulo 2.

We now understand a general method used again and again by Minsky and Papert: Start with a pattern-classification problem. Observe that certain symmetries leave it invariant (for instance, for the parity problem [Is the number of active elements even or odd?], which includes the case of addition modulo 2, any permutation of the points of the retina would leave the classification unchanged.) Use this to reduce the number of parameters describing the circuit. Then lump items together to get a polynomial and examine actual patterns to put a lower bound on the degree of the polynomial, fixing things so that this degree bounds the number of inputs to the response unit of a simple perceptron.

We now turn to a proof of this *group-invariance theorem*. Think of the retina R as having r elements, so that each preprocessor is defined by a *fixed* function $\phi: \{0, 1\}^r \to \{0, 1\}$.

1 Definition. For any collection Φ of functions $\phi: \{0, 1\}^r \to \{0, 1\}$ we define $L(\Phi)$, *the class of functions linear with respect to* Φ to be precisely those functions ψ that may be written

$$\psi = \left\lceil \sum_{\phi \in \Phi} \alpha_\phi \phi > \theta \right\rceil;$$

that is, $\psi(x) = 1$ iff it is true that $\sum \alpha_\phi \phi(x) > \theta$ for suitable choices of real numbers α_ϕ (the "weights") and θ (the "threshold").

In other words, if Φ is the set of preprocessor functions for a simple perceptron, then, no matter how we adjust the weights of the response unit, the response $\{0, 1\}^r \to \{0, 1\}$ of the perceptron must belong to $L(\Phi)$. We shall say a function ϕ of Φ is of *degree k* if we may associate it with a neuron having k input lines, each a distinct line from the retina R. We then say that the *order* of ψ is the smallest integer k for which $\psi \in L(\Phi)$ for some collection Φ of functions in which every ϕ is of degree $\leq k$. Thus, a linear threshold function is of order 1, and every function is of order $\leq |R|$ (where we now use $|S|$ to denote the number of elements in the set S, so that $r = |R|$ is the size of the retina). Minsky and Papert ask how big an order is required for a function if we bound the depth of the network by only allowing one level read out by a threshold element.

2 Definition. ϕ is called a mask, and written ϕ_A, iff there is a set A such that $\phi(X) = \lceil A \subset X \rceil$. We write ϕ_x for $\phi_{\{x\}}(X) = \lceil x \in X \rceil$. Let μ_R be the set of all masks on R.

Thus, ϕ_A is of *degree* $|A|$ and is simply an AND gate with one input line for each element of A. On the other hand, since $\phi_A = \lceil \sum_{x \in A} \phi_x \geq |A| \rceil$, we have that the *order* of every mask is 1.

3 Theorem. *Every Boolean function* ψ *is a linear threshold function with respect to the set of all masks; that is, every* ψ *is in* $L(\mu_R)$.

PROOF. For a binary variable x, let us use the notation x^0 for $\bar{x} = 1 - x$, and x^1 for x itself. Then for n variables $x_1, \ldots, x_n$ and for fixed binary values $\alpha_1, \ldots, \alpha_n$, the expression

$$x_1^{\alpha_1} x_2^{\alpha_2} \cdots x_n^{\alpha_n}$$

(where multiplication replaces conjunction) has the interesting property that it equals 1 if and only if $x_1 = \alpha_1$, $x_2 = \alpha_2$, ... and $x_n = \alpha_n$. But this immediately implies that our arbitrary Boolean function ψ of the n variables x_1, x_2, ..., x_n satisfies the equation

$$\psi(x_1, \ldots, x_n) = \sum \psi(\alpha_1, \ldots, \alpha_n) x_1^{\alpha_1} \cdots x_n^{\alpha_n}.$$

The right-hand side is called the *disjunctive normal form* for ψ.

Gathering terms this becomes

$$\psi(x) = \sum \gamma_i \phi_i(X),$$

where each $\phi_i(X) = x_{j_1} \cdots x_{j_m}$ for some subset $(j_1, \ldots, j_m)$ of $(1, \ldots, n)$, which just says ϕ_i is a mask. Thus

$$\psi = \lceil \sum \gamma_i \phi_i > 0 \rceil,$$

so that ψ does indeed belong to $L(\mu_R)$.   □

We shall now show that if a predicate is unchanged by various permutations, then we may use this fact to simplify its coefficients with respect to the set of masks, and that this simplified form will often enable us to place a lower bound on the order of the predicate.

Let G be a group of permutations on R, in other words, G is simply a rearrangement of R. We write xg for the image of $x \in R$ under $g \in G$. Thus, for each $x_1, x_2, \in R$, we have $x_1 g \neq x_2 g$ if $x_1 \neq x_2$; while for every $y \in R$ there is an $x \in R$ (which must be unique) such that $xg = y$. We then write for $x \subset R$

$$XG = \{xg \mid x \in X \text{ and } g \in G\}$$

the set of all points reachable from points in X by applying transformations from G. We use ϕg to denote the function with $\phi g(X) = \phi(Xg)$. We say ψ is *invariant under G* just in case $\psi = \psi g$ for all $g \in G$.

We then say ϕ is equivalent to ϕ' with respect to G, and write $\phi \equiv_G \phi'$, just in case $\phi = \phi' g$ for some $g \in G$, i.e., ϕ and ϕ' only differ by the relabeling of their arguments x by some permutation g from G. A *block* of $\equiv_G$ is simply a set of all the ϕ equivalent to one another, i.e., it is an equivalence class with respect to $\equiv_G$.

4 The Group Invariance Theorem. *Let G be a group of permutations of R and let* Φ *be a set of functions on R closed under G (that is, $\phi \in \Phi$, $g \in G$ implies $\phi g \in \Phi$). Then if ψ in $L(\Phi)$ is invariant under G, so that $\psi g = \psi$ for every g in G, it has a linear representation*

$$\psi = \left\lceil \sum_{\phi \in \Phi} \beta(\phi)\phi > \theta \right\rceil,$$

in which $\beta(\phi) = \beta(\phi')$ whenever $\phi \equiv_G \phi'$.

PROOF. Given a representation

$$\psi = \lceil \sum \alpha(\phi)\phi > \theta \rceil$$

form

$$\beta(\phi) = \sum_{g \in G} \alpha(\phi g)/|G|,$$

which thus depends only on the equivalence class of ϕ. Then $\psi = \lceil \sum \beta(\phi)\phi > \theta \rceil$, as the reader may readily verify. □

5 Corollary. *Let* $\Phi = \Phi_1 \cup \cdots \cup \Phi_m$, *where each* Φ_i *is a block of* $\equiv_G$. *Let* $N_i(X)$ *be the number of* ϕ's *in* Φ_i *for which* $\phi(X)$ *is true. Then if* ψ *is in* $L(\Phi)$, *with* Φ *closed, and* ψ *invariant, under* G, *then* ψ *has a representation*

$$\psi = \sum_{i=1}^{m} \alpha_i N_i > \theta.$$

We may now apply the group invariance theorem to show that the order of the parity function equals $|R|$. In other words, parity is so odd that the Boolean preprocessors must be complex enough to do all the work, rendering the response unit superfluous.

6 Theorem. *The parity function*

$$\psi_{PAR}(X) = \lceil |X| \text{ is an odd number} \rceil$$

is of order $|R|$.

PROOF. Since ψ_{PAR} is invariant under the group G of all permutations of R, the Corollary tells us that ψ_{PAR} has a representation

$$\psi_{PAR} = \left\lceil \sum_{\alpha} \alpha_j C_j > \theta \right\rceil,$$

where $C_j(X)$ is the number of masks ϕ of degree j with $\phi(X) = 1$ and thus equals the number of subsets of X with j elements:

$$C_j(X) = \binom{|X|}{j} = \frac{1}{j!} |X|(|X|-1)\cdots(|X|-j+1),$$

a polynomial of degree j in $|X|$.

If ψ_{PAR} is of order K, then $P(X) = \sum_{j=0}^{k} \alpha_j C_j(X) - \theta$ is a polynomial of degree $\leq K$ in $|X|$.

Now, let X_j have j points, $j = 0, 1, \ldots, |R|$. Then the sequence $P(|X_0|) \leq 0$, $P(|X_1|) > 0, P(|X_2|) \leq 0, \ldots, P(|X_R|)$ changes sign $|R|-1$ times. Thus, P has degree $\geq |R|$, and so we conclude that ψ_{PAR} must have order $|R|$. □

Thus, Minsky and Papert have been able to show in a precise numerical formula how the complexity of the components in the first layer of a two-layer network must increase with the complexity of the pattern-recognition problem required, in this case recognition of parity for patterns of increasing size. To tell whether the number of squares that are on is even requires neurons that actually are connected to all of the squares of the network. By contrast, to tell whether the number of on squares reaches a certain threshold only requires two inputs per neuron in the first layer:

7 Proposition. *If M is an integer $0 < M < |R|$, then the "counting function" $\psi^M(X) = \lceil |X| = M \rceil$ is of order ≤ 2.*

PROOF.

$$\psi^M(X) = \lceil (|X| - M)^2 \leq 0 \rceil$$

$$= \lceil (2M - 1)|X| - 2|X|(|X| - 1) \geq M^2 \rceil$$

$$= \lceil (2M - 1) \sum_x \phi_x(X) + (-2) \sum_{x \neq x'} \phi_{\{x,x'\}}(x) \geq M^2 \rceil. \qquad \square$$

A further application of the group-invariance theorem shows that to tell whether or not the pattern of activated squares is connected requires a number that increases at least as fast as the square root of the number of cells in the retina. These results are most interesting and point the way toward further insight into the functioning of the nervous system, but they are restricted to highly mathematical functions instead of the complex perceptual problems involved in the everday life of an organism. We might note, too, that any full model of perception must not have the purely passive character of the perceptron model, but must involve an active component in which hypothesis formation is shaped by the inner activity of the organism and related to past and present behavior [Craik, 1943; Gregory, 1969].

The Winograd–Spira theory

We have already stressed the interest, in the study of complexity of computation, in tradeoffs between time and space. Minsky and Papert asked, "If we fix the number of layers in the network, how many input lines must the elements have in order to get a successful computation?" We close this section by noting that Winograd and Spira have tackled the complementary problem of how, if we bound the number of inputs per component, we can proceed to discover how many layers of components we require. More specifically, they studied algebraic functions (e.g., group multiplication) instead of the problem of classifying patterns. Winograd, 1967, and later Spira and Arbib, 1967, and Spira, 1969, studied networks whose components were limited in that there was a fixed bound on the number of input lines to any component. In what follows, each module is limited to having at most r input lines. We are once again assuming a unit delay in the operation of all our modules.

The Winograd–Spira theory is based on the simple observation exemplified by Figure 4.16. Here we see that if there are two inputs per module and if an output line of a circuit depends on 2^3 input lines, then it takes at least three

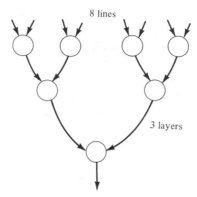

Figure 4.16   With fan-in of 2 lines per cell, at most eight input lines can affect the output of a cell at depth 3 of the network.

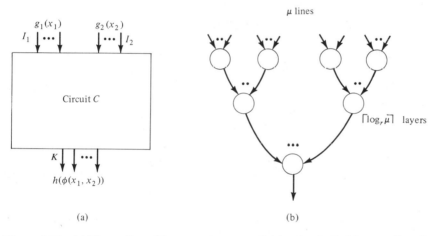

(a)   (b)

Figure 4.17   (a) The coding of inputs and outputs for the circuit C; (b) generalizes the situation of Figure 4.16 for μ input lines to a circuit with fan-in of r lines per cell.

time units for an input configuration to yield its corresponding output. A lemma below formalizes this observation, and is the basis for the lower bound we obtain on computation time for various functions.

Spira and Arbib, 1967, made explicit the methodology implicit in Winograd's work, forming a basis upon which we can erect a thoroughgoing analysis of multiplication in groups and semigroups and can also analyze computation of various finite functions. We present here part of such an analysis. Our concern is with a circuit C [Figure 4.17(a)] such that we can code $x_1 \in X_1$ as an input configuration $g_1(x_1)$ to C, and could eventually have C compute $\phi(x_1, x_2)$ by emitting a coded version $h(\phi(x_{ij}x_2))$.

8 Definition. Let $\phi: X_1 \times X_2 \to Y$ be a function of finite sets. Let circuit C have input lines partitioned into two sets with I_j the set of possible configurations on the jth ($j = 1, 2$) set of input lines, and K the set of possible output configurations; we say C is capable of computing the function ϕ in time τ if there is a state s_0 of C, maps $g_j: X_j \to I_j$ ($j = 1, 2$), and a one-to-one function $h: Y \to K$ such that if C is started in state s_0 at time 0 and receives constant input $(g_1(x_1), g_2(x_2))$ from time 0 through time $\tau - 1$, the output at time τ will be $h(\phi(x_1, x_2))$.

9 Lemma. *In a circuit with at most r inputs per module, the output of an element at time τ can depend upon preceding values on at most r^τ input lines.*

PROOF. Just consider the fan-in with modules having r input lines each to the height of τ [see Figure 4.17(b)].   □

Changing our earlier notation, we now let $\lceil x \rceil$ be the smallest integer $\geq x$; $\lfloor x \rfloor$ the largest integer $\leq x$. Let $h_j(y)$ be the value on the jth output line when the overall output configuration is $h(y)$.

10 Definition. Let $\phi: X_1 \times X_2 \to Y$ and let $h: Y \to K$ be an output function for a circuit C which computes ϕ. Then $S \subset X_1$ is called *an h_j separable set for C in the first argument of ϕ* if s_1, $s_2 \in S$ and $s_1 \neq s_2$ implies that there exists $x_2 \in X_2$, with

$$h_j(\phi(s_1, x_2)) \neq h_j(\phi(s_2, x_2)).$$

Similarly for sets separable in the second argument.

This notion of h_j-separable set allows us to use the fan-in lemma to get a completely general lower bound on computation time.

11 The Basic Lemma. *Let $\phi: X_1 \times X_2 \to Y$. Let C be a circuit with fan-in r which computes ϕ in time τ. Then*

$$\tau \geq \max_j \lceil \log_r (\lceil \log_2 |S_1(j)| \rceil + \lceil \log_2 |S_2(j)| \rceil) \rceil$$

where $S_i(j)$ is an h_j-separable set for C in the ith argument of ϕ.

PROOF. The jth output at time τ must depend on at least $\lceil \log_2 |S_i(j)| \rceil$ output lines from I_i or else there would be two elements of $S_i(j)$ that were not h_j separable. Thus, the jth output depends on at least $\lceil \log_2 |S_1(j)| \rceil + \lceil \log_2 |S_2(j)| \rceil$ input lines, from which $r^\tau \geq \lceil \log_2 |S_1(j)| \rceil + \lceil \log_2 |S_2(j)| \rceil$ and the result follows, since τ is integral valued.   □

With the basic lemma we have exposed the methodology implicit in Winograd's treatment of the times required for addition and multiplication. Let us show its usefulness by giving two examples, where $U_N = \{0, 1, \ldots, N - 1\}$.

12 Example. Let $\phi\colon U_N \times U_N \to \{0, 1\}$ be

$$\phi(x, y) = \begin{cases} 1 & \text{if } x \le y, \\ 0 & \text{if } x > y. \end{cases}$$

Then if the circuit C with fan-in r computes ϕ in time τ, we have

$$\tau \ge \lceil \log_r(2\lceil \log_2 N \rceil) \rceil.$$

PROOF. If $h_j(0) \ne h_j(1)$, then U_N is an h_j-separable set for C in both the first and second arguments of ϕ, since if $x > y$, $\phi(x, y) \ne \phi(y, y)$ and $\phi(x, y) \ne \phi(x, x)$. $\qquad\square$

13 Example. Let $\phi\colon U_N \times U_N \to U_N$ be defined by $\phi(x_1, x_2) = \lfloor x_1 \cdot x_2/N \rfloor$. Then, if C computes ϕ in time τ, we have

$$\tau \ge \lceil \log_r(2\lceil \log_2 \lfloor N^{1/2} \rfloor \rceil) \rceil.$$

PROOF. Pick j such that $h_j(0) \ne h_j(1)$. Let $m = \lfloor N^{1/2} \rfloor$. Then $\{1, 2, \dots, m\}$ is an h_j-separable set for C in both arguments of ϕ, since for each $x \ne y$ with $x, y \in 1$, $2, \dots, m$ we may choose $z \in U_N$ to be such that $x \cdot z < N \le y \cdot z < 2N$ to yield $\phi(x, z) \ne \phi(y, z)$. By symmetry this holds for the second argument as well, and the result follows. $\qquad\square$

Winograd and Spira use such methods to prove for certain mathematical functions that no possible scheme of wiring neurons can compute the function in less than a certain time delay, intimately related to the structure of the function. In particular, they showed how to go from the structure of a finite group to a minimal time for a network computing the group multiplication. Furthermore, Spira, 1969, was able to provide for any group a network that is essentially time-optimal, in that it produces its output within one time unit of the time specified by the theorem.

Implications

We thus have an extremely exportant result for any theory of neural networks: for a certain type of restricted component we can show how to build a network that is optimal with respect to the time required for computing. However, to appreciate the full complexity that lies ahead of the automata theorist who would contribute to the study of information processing in the nervous system, we must make several observations. First, to achieve time optimality in his network, Spira has to use an extremely redundant encoding for the input and output to ensure that "the right information would be in the right place at the right time." The flavor of this can be given by the observation that we can multipy numbers far more quickly if they are given in prime decomposition, for example,

$$(2^2 \cdot 3^4 \cdot 5^2 \cdot 7^1) \times (2^1 \cdot 3^0 \cdot 5^1 \cdot 7^2) = (2^3 \cdot 3^4 \cdot 5^3 \cdot 7^3),$$

than if they are given in decimal form, for in the former case we need only add exponents $(2 + 1 = 3, \ 4 + 0 = 4, \ 2 + 1 = 3, \ 1 + 2 = 3)$ instead of going through the lengthy computation of decimal multiplication.

This observation reminds us of the organization of the cat visual system, where a million optical fibers feed 540 million cells in the visual cortex. Thus, as we move up into the visual cortex, what we reduce is not the number of channels, but rather the activity of the channels, as each will respond only to more and more specific stimuli (Barlow, 1969). The result is a network with many, many neurons in parallel, even though the network is rather shallow in terms of computation time. It might well be that we could save many neurons, at the price of increased computation time, both by narrowing the net while increasing its depth and by using feedback to allow recirculation of information for a long time before the correct result emerges. We see here the need for a critical investigation of the interplay between space and time in the design of networks.

The ganglion cells in the retina of a frog seem fairly well suited to a life spent living in ponds and feeding on flies, since the brain of the animal receives specific information about the presence of food and enemies within the visual field (Lettvin et al., 1959). But the price the animal pays is that it is limited in its flexibility of response because its information is so directly coded. A cat (Hubel and Wiesel, 1962), on the other hand, has to process a greater amount of information if it is to find its prey, but it can eat mice instead of flies. A cat cannot compute its appropriate action as quickly, perhaps, as the frog can, but it makes up for that by having extra computational machinery enabling it to predict and make use of previous experience in developing a strategy for governing its action.

We see that to model the behavior of the animal completely, we must make an adequate model of its environment and take into account structural features of the animal. It is not enough to work out an optimum network whereby a frog can locate a fly; we must also compute whether it is optimal to couple that network to the frog's tongue, or to have the frog bat the fly out of the air with its forelimb, or to have the frog jump up to catch the fly in its mouth. Clearly the evolution of receptors, effectors, and central computing machinery was completely interwoven, and it is only for simplicity of analysis that we concentrate here on the computational aspects, holding many of the environmental and effector parameters fixed. Again, we shall ignore the interesting pattern-recognition problem of determining the most effective features to be used in characterizing a certain object in a given environment. For instance, to characterize a mouse one could go into many details including the placement of hairs upon its back, but for the cat it is perhaps enough to recognize a gray or brown mobile object with a tail and within a certain size range. It should be clear that the choice of features must depend on the environment; if there exists a creature that meets the above prescription for a mouse but happen to be poisonous, then it will clearly be necessary for a successful species of cat to have a perceptual system that can detect features differentiating the poisonous creatures from the edible mice.

We should further note that Winograd and Spira's best results were for groups, where we can make use of the mathematical theory elaborated over the past hundred years. It will be much harder to prove equally valuable theorems about functions that are not related to classical mathematical structures. If we were to replace the very simple limitation on number of inputs by an assumption limiting the actual types of functions that the neurons can compute, we would have to expect a great increase in the complexity of the theory. Some would conclude from this analysis that automata theory is irrelevant to the study of the nervous system, but we would argue that it shows how determined our study of automata theory must be before we can hope to understand fully the function of the nervous system.

References for Chapter 4

Amari, S., 1980, Topographic organization of nerve fields, *Bull. Math. Biol.* **42**: 339–364.

Amari, S. and Arbib, M.A., 1977, Competition and cooperation in neural nets, in *Systems Neuroscience* (J. Metzler, Ed.), Academic Press, pp. 119–165.

Barlow, H.B., 1953, Summation and inhibition in the frog's retina, *J. Physiol.* **119**: 69–88.

Barlow, H.B., 1969, Trigger features, adaptation and economy of impulses, in *Information Processing in the Nervous System* (K.N. Leibovic, Ed.), Springer-Verlag, pp. 209–230.

Barnes, D.B., 1986a, From genes to cognition, *Science* **231**: 1066–1068.

Barnes, D.B., 1986b, Lessons from snails and other models, *Science* **231**: 1246–1249.

Blakemore, C. and Cooper, G. 1970, Development of the brain depends on the visual environment, *Nature* **228**: 477–478.

Block, H.D., 1962, The perceptron: A model for brain functioning, 1. *Rev. Mod. Phys.* **34**: 123–135.

Churchland, P.S., 1986, *Neurophilosophy: Toward a Unified Science of the Mind/Brain*, A Bradford Book/The MIT Press.

Cooper, L.N., Liberman, F., and Oja, E., 1979, A theory for the acquisition and loss of neuron specificity in visual cortex, *Biological Cybernetics* **33**: 9–28.

Craik, K.J.W., 1943, *The Nature of Explanation*, Cambridge University Press.

Didday, R. L., 1970, The simulation and modelling of distributed information processing in the frog visual system. Ph.D. Thesis, Stanford University.

Fregnac, Y. and Imbert, M., 1978, Early development of visual cortical cells in normal and dark-reared kittens: Relationship between orientation selectivity and ocular dominance, *J. Physiol. London* **278**: 27–44.

Fukushima, N., 1980, Neocognitron: A self-organizing neural network model for a mechanism of pattern recognition unaffected by shift in position, *Biological Cybernetics* **36**: 193–202.

Gregory, R.L., 1969, On how so little information controls so much behavior, in *Towards a Theoretical Biology. 2, Sketches* (C.H. Waddington, Ed.), Edinburgh University Press.

Grossberg, S., 1982, *Studies of Mind and Brain*, Reidel.

Hartline, H.K., 1938, The response of single optic nerve fibers of the vertebrate eye to illumination of the retina, *Am. J. Physiol.* **121**: 400–415.

Hirsch, H., and Spinelli, D.N., 1970, Visual experience modifies distribution of horizontally and vertically oriented receptive fields in cats, *Science* **168**: 869–871.

Hubel, D.H., and Wiesel, T.N., 1962, Receptive fields, binocular and functional architecture in the cat's visual cortex. *J. Physiol.* **160**: 106–154.

Hubel, D.H., and Wiesel, T.N., 1965, Binocular interaction in striate cortex of kittens reared with artificial squint, *J. Neurophysiol.* **28**: 1041–1059.

Kandel, E., 1978, *A Cell Biological Approach to Learning*, Grass Lecture No. 1, Society for Neuroscience.

Kilmer, W.L., and Olinski, M., 1974, Model of a plausible learning scheme for CA3 hippocampus, *Kybernetik* **16**: 133–144.

Kohonen, T., 1982, in *Competition and Cooperation in Neural Nets* (S. Amari and M.A. Arbib, Eds.), *Lecture Notes in Biomathematics*, Vol. 45, Springer-Verlag.

Lara, R., 1983, A model of the neural mechanisms responsible for stimulus specific habituation of the orienting reflex in vertebrates, *Cognition and Brain Theory* **6**: 463–482.

Lettvin, J.Y., Maturana, H., McCulloch, W.S., and Pitts, W.H., 1959, What the frog's eye tells the frog's brain, *Proc. IRE.* 1940–1951.

Milner, P.M., 1957, The cell assembly: Mark II, *Psych. Rev.* **64**: 242.

Minsky, M.L., and Papert, S., 1969, *Perceptrons: An Essay in Computational Geometry*, The MIT Press.

Nilsson, N., 1965, *Learning Machines*, McGraw-Hill.

Pitts, W.H., and McCulloch, W.S., 1947, How we know universals, the perception of auditory and visual forms. *Bull. Math. Biophys.* **9**: 127–147.

Pribram, K.H., Spinelli, D.N., and Kamback, M.C., 1967, Electrocortical correlates of stimulus response and reinforcement, *Science* **157**: 94–96.

Rosenblatt, F., 1962, *Principles of Neurodynamics*, Spartan.

Rumelhart, D.E., and Zipser, D., 1985, Feature discovery by competitive learning, *Cognitive Science* **9**: 75–112.

Sklansky, J., and Wassel, G.N., 1981, *Pattern Classifiers and Trainable Machines*, Springer-Verlag.

Sokolov, E., 1975, Neuronal mechanisms of the orienting reflex, in *Neuronal Mechanisms of the Orienting Reflex* (E. Sokolov and E. Vinogradova, Eds.), Lawrence Erlbaum Associates.

Spinelli, D.N., 1970, OCCAM, A computer model for a content addressable memory in the central nervous system, in *Biology of Memory* (K.H. Pribram and D.E. Broadbent, Eds.), Academic Press, pp. 293–306.

Spinelli, D.B., and Jensen, F.E., 1979, The mirror of experience, *Science* **203**: 75–78.

Spira, P.M., 1969, The time required for group multiplication, *J. Assoc. Comp. Mach.* **16**: 235–243.

Spira, P.M., and Arbib, M.A., 1967, Computation times for finite groups, semigroups and automata, *Proc. IEEE 8th Ann. Symp. Switching and Automata Theory*, pp. 291–295.

von der Malsburg, C., 1973, Self-organizing of orientation sensitive cells in the striate cortex. *Kybernetik* **14**: 85–100.

von der Malsburg, C., 1979, Development of ocularity domains and growth behavior of axon terminals, *Biological Cybernetics* **32**: 49–62.

Winograd, S., 1967, On the time required to perform multiplication, *J. Assoc. Comp. Mach.* **14**: 793–802.

CHAPTER 5
Learning Networks

5.1 Connectionism

The cognitive science that emerged in the 1970s was based mainly on the serial information processing paradigm of artificial intelligence (AI) and the symbol-manipulation approach to linguistics, and had rather little contact with work in brain theory. However, there is now a growing interest in what is called the connectionist approach or Parallel Distributed Processing (PDP), the study of ways in which simple units may be interconnected to solve hard problems. This approach may to some extent be characterized as a reaction against the domination of AI by the paradigms of serial computation and, in some cases, explicit symbolic structures; but it can also be seen as the result of probing the microstructure of "symbols" and thus stressing the parallel processes underlying behavior. Of course, to the reader of this book, the approach is also a direct continuation of the work on adaptive pattern-recognition networks that we have sampled in Chapter 4.

This paradigm is, as we have seen, informed by analogies with neural networks, rather than by the attempt to model real neural structures in any detail. To make an important distinction: Neural modeling is concerned with modeling specific neural circuitry in real brains, and thus its theories eventually stand or fall to the extent that they make contact with experimental data about the brain. Connectionism (let us use one word, for simplicity) may be called *neurally inspired* modeling, or modeling *in the style of the brain*. Here the goal is to abstract the principles of computation employed by the brain: What kind of computer is it? How does it work? How can it be modified? How does it modify itself? Clearly, such an effort may be conducted as part of brain theory, constructing a conceptual framework within which to erect specific models linked to the data of neuroscience. It may also be part of an effort to find ways of building intelligent machines (AI) or of modeling the human mind, in which case the "reality constraints" are provided by the data of psychology.

An example of the way in which connections may constrain a set of partial solutions in solving an overall problem is given in Figure 5.1 from Rumelhart and McClelland (1986a). In Figure 5.1(a), we see a Necker cube, each vertex of which stimulates those nodes in the upper network that might be possible interpretations of the vertex. Excitatory connections between nodes express

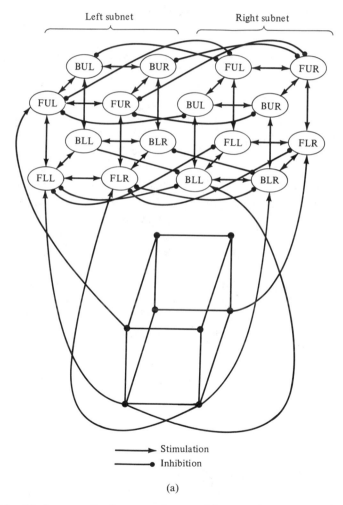

(a)

Figure 5.1 (a) A connectionist network comprising two interconnected networks, one for each interpretation of the Necker cube: B = back; F = front; R = right; L = left; U = upper, L = lower. (b) Three runs of the network. (After Rumelhart and McClelland, 1986a.)

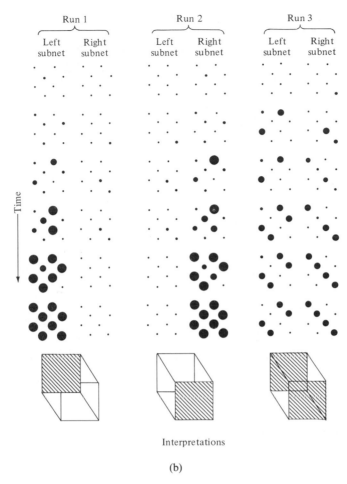

Interpretations

(b)

Figure 5.1   (*Continued*)

the likelihood that the two nodes could represent adjacent vertices in a coherent view of the cube, while inhibitory connections link mutually inconsistent hypotheses. Figure 5.1(b) shows three runs of the network. Each unit's activation level indicates the current "confidence level" of the hypotheses that what the unit stands for is present in the perceptual input. In two of the runs, the initial weak activation of many vertices yields a strongly held interpretation, while the third run is an example of an anomalous interpretation.

Connectionism may be characterized by

(a) The use of networks of active computing elements, with programs residing in the structure of interconnections.

(b) The restriction of these elements to a simple structure. The proper structure for such a unit is evolving, but the element is more like a linear threshold unit or a McCulloch–Pitts neuron (perhaps augmented by a few numerical state variables and a noise term) than either a real neuron or a computer program of moderate complexity.

(c) Massive parallelism, with no centralized control (other than control that can be exerted by other massively parallel nets).

(d) The encoding of semantic units either by single network units or by patterns of activity in a population of such units.

A major thrust has been the study of "learning algorithms" to generate a "useful" network through changing the connections between units. Such learning algorithms serve in two distinct roles: as psychological models of human learning or cognitive development and as means to "program" nets to satisfy given performance specifications.

The networks studied in this way have certain important features:

(a) Ability to bring multiple interacting constraints to bear on problem-solving.

(b) Associative and content-addressable memory.

(c) Generalization: adaptability in parametrized systems in which the topology of parameter space can be linked to similarity of the functions that the networks are to serve.

(d) Rules as emergent properties rather than explicit symbolic structures.

(e) Speed of processing from exploitation of parallelism.

These properties are to result from general architectural principles, rather than being built in one at a time. Among these general principles are

(1) Cooperative computation.

(2) Distributed representations/coarse coding.

(3) Relaxation and constraint satisfaction algorithms.

As we shall see below, the language of energy landscapes sometimes provides a useful tool for the analysis of network properties.

We may recognize a dichotomy in the structure of neuronlike networks. In one case, *localized* representations are used, in which each neuron represents a concept, much as if a semantic network were turned into a neural net. Such a representation is transparent and easy to implement. In a *distributed* representation, on the other hand, symbols are coded by patterns, thus enforcing an associational structure in which symbols coded by similar patterns tend to have similar associations. Such a representation may be the natural outcome of a learning procedure "letting the features fall where they may." In general, this distribution may be quasilocal in that a representation will be distributed across the neurons of some region of the net, rather than across the network in its entirety.

In summary, note that connectionism is not synonymous with "use of distributed computation in AI," for this designation may include systems in which several large systems do a great deal of computation internally, and then communicate via large symbol structures. By contrast, in a connectionist network, the units are small, each doing a small part of the computation in parallel, and communicating via activation levels. However, my own opinion is that these two approaches are not so much alternatives as complementary. Certainly, in brain theory, the ultimate analysis is in terms of neurons, but first we seek structural units (e.g., anatomically defined brain regions) or functional units (schemas) of intermediate complexity, and it is these units that we map (perhaps again via intermediate structures such as modules or neural layers) to actual neural networks. (It should be noted, however, that if different schemas are mapped onto activity in the same set of units, then the implementation may affect the functioning of the schema.) This approach is that taken in the companion volume *The Metaphorical Brain*, Second Edition.

With this, let us briefly continue the historical analysis of Chapter 1, tracing some of the ways in which a concern with parallelism and distributed processing began to erode the serial paradigm in AI, starting in the mid-1970s. By this time, workers in AI had come to realize the complexity of intelligent problem-solving in such fields as language understanding and vision. Explicit formalisms for semantic nets, scripts, frames, and other forms of knowledge representation were developed in response to a consensus that intelligence was based not only on processing capabilities but also on access to a vast amount of detailed information, and that there was seldom an easy way to access that information without a great deal of search. Meanwhile, work on constraint satisfaction and relaxation algorithms made it clear that such search could often be conducted in parallel networks. For example, Waltz (1975) showed how to interpret a line drawing as a set of blocks by using a network in which each node of the net corresponded to a vertex of the drawing, and the links between nodes embodied the crucial knowledge as to which pairs of interpretations were possible for these vertices. Matching elements from a stereo pair to infer the depth of objects in the original scene and cleaning up edge elements to fit smooth curves were other problems that Rosenfeld, Hummel, and Zucker, 1976, and Hummel and Zucker, 1983, addressed in a unified methodology of constraint propagation, or relaxation, in networks.

Such developments meant increasing congruence with earlier work in pattern recognition and brain theory. Selfridge, 1959, had developed the Pandemonium model of pattern recognition, in which a "chief demon" would listen to the "shouts" of "lesser demons," each specialized to emit a "howl" expressing their degree of confidence in the particular pattern they were specialized to recognize. By contrast, Kilmer, McCulloch, and Blum (1969) offered a network in which each module (itself a neural network), on the basis of its own input sample, formed initial estimates of the likelihood of each of a finite set of modes, and then communicated back and forth to achieve a consensus as to which mode was most appropriate for the given sample. This style of decision-making, involving communication among a net of modules rather than the imposition of the decision by a global controller may be called *cooperative computation.*

The HEARSAY model of speech understanding (Erman et al., 1980) made a major impact in developing the cooperative computation style in the AI community, for it encouraged the use of an architecture in which many independent knowledge sources were at work on a common "blackboard," and it was through their efforts that a coherent interpretation of some utterance was obtained. Although the actual implementation used a serial scheduler of activation of these computing agents (Arbib and Caplan, 1979, described how a "neurologized" HEARSAY might be structured), the essential logic of the system was that of cooperative computation, and encouraged many people to think of perception and language in these terms. Working with Arbib, Helen Gigley (1983) developed a "lesionable" model of parsing, designed to address data on aphasia, based on competition and cooperation in neural nets; at about the same time, as members of the Rochester school, Gary Cottrell and Steven Small (1983) developed a connectionist approach to the semantics of natural language, based on case-frame semantics, while a related study is due to Waltz and Pollack, 1985. Another example of the influence of HEARSAY, this time in cognitive psychology, is offered by the work of Rumelhart. He started from mathematical learning theory, but became interested in semantic nets. However, these were mainly serial and looking at perception he knew that Pandemonium was more nearly correct. He wrote a paper in 1975 telling psychologists to pay attention to HEARSAY and cooperative computation. He tried to implement HEARSAY with schemas as demons, but the resulting system was too complicated. The attempt to simplify led to the work with McClelland (McClelland and Rumelhart, 1981; Rumelhart and McClelland, 1982) on the interactive activation model of reading (Figure 5.2), in which computation was based on the parallel interaction of many small units, rather than the scheduled interaction of large units as in HEARSAY.

Another element in modern connectionism stems from coming to view a computer not in terms of a forced lockstep through a sequence of discrete states but rather in terms of the motion of a dynamic system toward some equilibrium or attractor set. This concept was part of the initial impetus for cybernetics: Cannon's (1939) notion of homeostasis as a key element in metabolic regulation had a major influence that Wiener, 1948, acknowledges in the historical introduction to *Cybernetics*, while stability was the goal of Ashby's 1954 homeostat in *Design for a Brain*. (Admittedly, being in a stable state is rather boring: what is interesting is when a system has the resources to achieve such stability despite major perturbations, or the imposition of major obstacles.) Such ideas have always been current in theoretical biology. To pick just one example, C.H. Waddington related the ability of embryos to regulate, i.e., to yield a recognizable adult form despite many traumas, to what he called *homeorhesis* in his book *The Strategy of the Genes*. Here the notion is not that of moving toward a fixed equilibrium but rather of returning toward some overall trajectory despite some major perturbation. These ideas were very influential on Reńe Thom in the development of his theory of morphogenesis and structural stability.

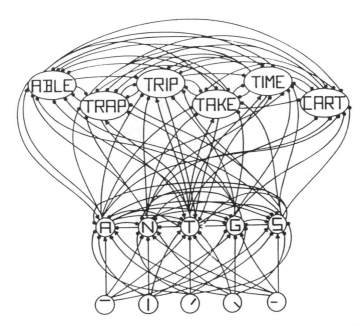

Figure 5.2 The connectionist network for word recognition of McClelland and Rumelhart, 1981, showing nodes for features, for *initial* letters of words, and for the words themselves. Note the inhibitory interactions between units for the initial letter, encouraging the net to give high activity to only one of these units ("winner takes all"), and similarly for the word units.

In some ways, then, a connectionist may see behavior as a series of settlings toward some attractors, a notion reminiscent of the Sherringtonian view of behavior as simply the chaining of reflex settlings into a series of postures. However, this very comparison reminds us that there are people working on movement who appeal directly to limit cycles and other properties of non-linear systems without recourse to connectionism or AI. One example is the work of Scott Kelso (cf. Kelso and Tuller, 1984), which can be traced back to the work of Peter Greene (1969) discussed in Arbib (1964).

Before turning to a detailed analysis of several connectionist models in the remaining sections of this chapter, we briefly introduce a few key players together with their contributions:

Jerry Feldman was the worker within the mainstream of AI who played the key role as proselytizer for massive parallelism. He listed constraints on processing by brains to achieve decisions in a few hundred milliseconds using components with a millisecond cycle time, and argued that we should exploit this parallelism in AI (Feldman, 1981). He is a prime exponent of the use of localist representations [cf. Feldman and Ballard (1982) for a manifesto]. Feldman espoused a "localist" style in which individual concepts are associated with individual nodes in the network. His colleague Dana Ballard provided connectionist networks for visual processing (Ballard and Brown, 1982, Ballard, Hinton, and Sejnowski, 1983), and has sought comparisons between connectionist networks and cerebral cortex (Ballard, 1986). Arnold Trehub has also provided connectionist solutions to explicit issues in using synaptic matrices in the design of a visual system [see Trehub (1987) for a review].

We should next mention three workers whose study of adaptive networks antedates its introduction to the AI community, yet who continue to make important contributions to the new rapprochement. James Anderson had worked on associative memory long before the current resurgence of connectionism, and is one of the few psychologists who really pushed the study of distributed memory systems. His brain-state-in-a-box (BSB) model (Section 5.2) is an important model of associative memory and has been used to model prototype effects, lexical access, and multistable perceptual phenomena. Teuvo Kohonen is a physicist who has also been working on associative memory for a long time, with good ties to mathematical analysis and engineering applications, and nice demonstrations of real applications. Stephen Grossberg (1970, 1978; Grossberg and Levine, 1975) is another persistent worker at the boundary between neural modeling and mathematical psychology and has been looking at adaptation algorithms via rigorous mathematics and dynamic systems for a long time. He has developed ingenious networks to solve a number of problems, including gain control. His work has more linkage to the behaviorist tradition of psychology than that of most other workers.

David Rumelhart and Jay McClelland are key players in the development of psychological explanations in terms of connectionist networks. They have shown how cognitive phenomena can emerge from patterns of interaction among elementary processing units, and they have just edited a book (Rumelhart and McClelland, 1986a) that provides an excellent introduction to PDP as a means for probing "the microstructure of cognition," showing how nets can have effects that are emergent and that would be unexpected in a symbol-processing domain. Their verb-learning paper (Rumelhart and McClelland, 1986b) is a compelling demonstration of "rule emergence," and does a good job empirically. Their interactive activation model explained the word-superiority effect for nonwords as the cumulative effect of partial matches with words (Figure 5.3). It was empirically successful, and is still accepted as the best explanation of the effect some six years later. The model predicted new effects, e.g., that SLNK has a word-superiority effect because of lots of partial word matches.

Andrew Barto has used reinforcement learning (cf. Section 5.4) to provide learning schemes for multilayered networks, addressing one of the major criticisms of learning networks by Minsky and Papert (Section 4.4), since it shows how to extend learning algorithms from single-layered networks to networks of greater complexity. The back-propagation algorithm (David

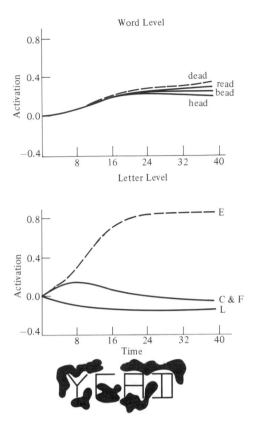

Figure 5.3   An example of nonword recognition using the network of Figure 5.2. Activation of letters for different positions can yield to activation of different word units that are in some sense similar to the displayed pattern (McClelland and Rumelhart, 1981). The display shows activity patterns for units linked to the second position in the word.

Rumelhart, Geoffrey Hinton, and Ron Williams, 1986; cf. Section 5.4) provides another approach to multilayered networks. Barto's invocation of stochastic learning theory and emphasis on a more cybernetic approach—adaptive control, not just pattern recognition—has initiated an important rapprochement of connectionism with engineering approaches to learning.

Geoffrey Hinton worked with Longuet-Higgins in the 1970s on relaxation methods, and played a catalytic role in the development of Rumelhart's ideas about PDP. He links studies in AI, cognitive psychology, and neuroscience. He has a key paper on implementing semantic networks using distributed representations/coarse coding [cf. Hinton and Anderson (1981, Chap. 6)].

John Hopfield (1982) viewed the dynamics of a symmetric network as minimizing a global energy measure [cf. Haken (1978) for earlier attempts to view the computation of a network in terms of energy/dynamical systems]. His paper served to interest many physicists in neural-like networks, although it added little to the understanding offered, e.g., by Anderson's BSB. His current push to implement such nets in silicon is also arousing interest. Hopfield views a net *with symmetrical connections* as a dynamical system operating according to some potential function. In other words, there are low-energy states and high-energy states. The state of the system changes as the state moves "down-hill" to a nearby energy minimum (cf. Figure 5.6 in Section 5.3). Hopfield uses *local* minima to store *different* patterns. Learning adjusts the potential surface so as to place the minima in desired locations. Where possible, he would like to shape the attractors, the "valleys" on the energy surface, too. The dynamics are somewhat like BSB dynamics, but here there are sigmoidal functions rather than hard limits.

Terry Sejnowski during the late 1970s examined the consequences of noise and random variability in the nervous system. Sejnowski, 1981, summarizes a formal stochastic analysis of nonlinear dynamic neural network models. The Boltzmann machine of Geoffrey Hinton and Terry Sejnowski (1983) extends Hopfield's nets (Section 5.3) by adding a "temperature"-dependent stochastic element that allows the use of simulated annealing. Just as a swordmaker anneals the metal, slowly reducing its temperature as he shapes it to ensure that the alloy is formed uniformly throughout the sword, so does simulated annealing lower a noise term, which is slowly reduced as the state of the network changes. This serves both as a technique for escaping local minima and as a way of mapping some extant mathematics onto a network (although simpler nets might achieve similar goals).

The work further illustrates the use of energy landscapes to view associative memories as settling into a fixed point in energy space. They add a learning rule with "hidden units" (i.e., even units that are not output units can be trained), but it is dependent on symmetric connections.

Rumelhart, Feldman, and Sejnowski (who by this time was working in neurobiology), together with such workers on associative memory as James Anderson (from the psychology perspective) and Teuvo Kohonen (from the pattern recognition/mathematical engineering perspective), were brought together at the UCSD conference published as Hinton and Anderson (1981). This book is perhaps the first full expression of connectionism as a new field that stretches across disciplinary boundaries. In the rest of this chapter, we briefly sample a number of recent studies of learning networks developed within the connectionist framework. Rather than following the style of de-tailed mathematical exposition that characterized, e.g., Sections 4.2 and 4.4, we here give a more journalistic exposition, designed to provide the reader with a useful set of entry points into a rapidly changing field. The reader who wishes to proceed further may turn to the excellent collection of articles written and edited by Rumelhart and McClelland (1986a), together with the article by Ballard (1986), which is published with a lively commentary and the author's response.

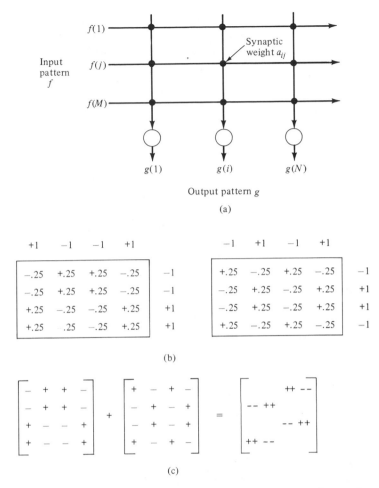

Figure 5.4 (a) Structure of the synaptic matrix. (b) Two synaptic matrices which ensure that the A pattern above the matrix will produce the B pattern at right. (c) The sum of these two matrices will, thanks to the orthogonality of the two A patterns in (b), reproduce both A-B pairings. (From Rumelhart and McClelland, 1986a.)

5.2 Synaptic Matrices

During the years in which AI was dominated by the serial computing paradigm, work continued in adaptive networks and associative memory. A number of authors consider networks [Figure 5.4(a)] in which there are m input lines each forming synapses on the same n neurons, so that the "state of learning" of the net is completely described by the *synaptic matrix* $A = [a_{ij}]$, where a_{ij} is the strength of the synapse from input i to neuron j. (For examples of this approach see the papers by Amari, Anderson, Kohonen, and Trehub.)

Synaptic learning rules are used that produce an *associative memory* which is *distributed* and thus *resistant to localized damage*. Little attempt is made to use a realistic neural model as the unit, e.g., many authors use a linear device. Some authors follow the McCulloch–Pitts convention in which neurons communicate via bits, 0 or 1, but in this section we look at models in which the activity levels are real numbers in $[-1, +1]$.

We speak of the pattern on the input lines as the *key* and the pattern on the output lines (the outputs of the neurons) as the *recollection*. In an *auto-associative* memory, learning is designed to make the key elicit the recollection of which it is a part, so that one has a content-addressable memory. In a *heteroassociative* memory, the key may be quite different from the recollection. In either case, what is restored for an arbitrary input is a linear combination of what has been stored weighted by the correlation of their keys with the current key. Life is simple if the keys are orthogonal; while Kohonen handles the general situation by use of the Moore–Penrose pseudoinverse. In auto-association, the response is the orthogonal projection of the key onto all the responses stored. (As the reader will see, the study of these networks requires a healthy familiarity with the tools of linear algebra. Kohonen (1984) offers this along with a sample of both technical and biological aspects of associative memory.)

Given a set f_k of keys and g_k of corresponding recollections, Anderson defines his synaptic matrix to be

$$A = \sum A_k \quad \text{where} \quad A_k = g_k f_k^T, \quad \text{i.e.,} \quad [A_k]_{ij} = g_{k(i)} f_{k(j)}. \tag{1}$$

Assuming that each scalar product $(f_j \cdot f_j)$ equals 1, we have that

$$Af_j = g_j + \sum_{k \neq j g_k} (f_k \cdot f_j) \tag{2}$$

which will equal g_j if the f_k's are orthonormal (Figure 5.4). If the keys are not orthogonal, one may alternatively use the Widrow–Hoff (1960) rule to form the associative mapping as a least means square solution.

One of Kohonen's most dramatic demonstrations is shown in Figure 5.5, a

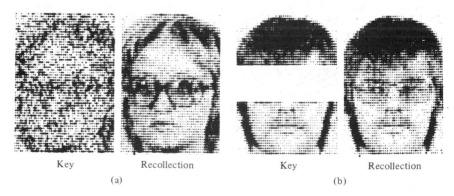

| Key | Recollection | Key | Recollection |
| (a) | (b) |

Figure 5.5   Demonstration of noise suppression and autoassociative recall in the orthogonal projection operation (Kohonen, 1977).

synaptic matrix that can store patterns each of which is an array of 3024 pixels (picture elements)—thus employing 3024 × 3024 synapses. This system could thus store 3024 orthogonal patterns, but Kohonen used it to store 100 arbitrary faces in a structure that is damage resistant, noise resistant, and content addressable. Perhaps this success is due to the sparseness of storing 100 patterns in almost 10,000,000 synapses—"but that's OK if synapses are cheap." This demonstration, although an important benchmark for the field, sidestepped the even more difficult problem of translational, rotational, and scale invariance, since all of the stored images were preregistered to the same position, orientation, and size.

Recurrent nets

Anderson's BSB (brain-states-in-a-box) model is an autoassociative synaptic matrix with the output fed back to the input. The response is a continuous (not a binary) function of the input, but with lower and upper saturation levels. What is learned is a desired tuple of extreme values, the weights being adjusted each time equilibrium is reached with a given input in such a way as to reduce the discrepancy with the desired value.

Anderson relates his model to human performance, with stabilization time being compared to human reaction times.

The learning rule for this recurrent net (Figure 5.6) is

$$a_{ij}(t + 1) = a_{ij}(t) + f(i)f(j), \tag{3}$$

as in the nonrecurrent case. Note that if $A(0)$ is symmetric, then $A(t)$ is

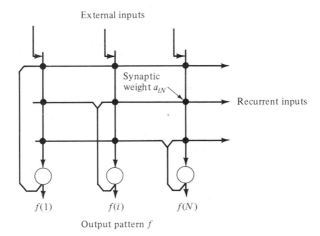

Figure 5.6   Recurrent net: A synaptic matrix [Figure 5.4(a)] but with the outputs fed back to provide the array of inputs; in addition, an external input is applied to each neuron.

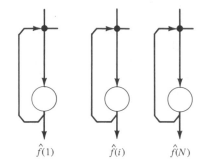

$$\hat{f}(1) \qquad\qquad \hat{f}(i) \qquad\qquad \hat{f}(N)$$

Figure 5.7   A decoupled recurrent net.

symmetric for all $t \geq 0$. Assume that $f_1, \ldots, f_k$ is an orthonormal set, and that f_i is presented k_j times as a forced firing pattern, then

$$A = \sum_{i=1}^{k} k_i f_i f_i^T. \tag{4}$$

If A is real and symmetric, there is a basis $e_1, \ldots, e_N$ consisting of orthogonal eigenvectors of A: $Ae_i = \beta_i e_i$ for $i = 1, \ldots, N$ with each β_i real. Now

$$Af_j = \sum_{i=1}^{k} k_i f_i f_i^T f_j = k_j f_i, \tag{5}$$

since the f_i's are an orthonormal set, and so the f_i's *are* the eigenvectors. If $k < N$, the remaining eigenvalues are 0, their eigenvectors complete the basis. If we were to change the basis to code f_i as $(0 \ldots 1 \ldots 0)$—with the 1 in the ith place—the net takes the simple form shown in Figure 5.7.

A is essentially a simple covariance matrix of the process generating the patterns. The eigenvectors of A correspond to "components," and those with the largest eigenvalues to "principal components" (cf. factor analysis). In this orthonormal case, the eigenvalue is just a measure of the number of presentations. The general case involves a "weighted combination" of eigenvectors. The use of eigenvalues allows one to discard the "infrequent" components to come up with vectors in a lower-dimensional space that over time provides a reasonable approximation as "averaged over time." What is encountered less frequently is less likely to be remembered. We thus require an entirely different type of model to analyze the human ability to remember certain events in "one shot."

Let $x(t)$ be the activity vector at time t, with $x(i, t)$ its ith component. Anderson posits no spontaneous decay in x, setting $x(t + 1) = x(t) + Ax(t) = (1 + A)x(t)$. If $x(t)$ is an eigenvector with eigenvalue β, then $x(i, t + 1) = (1 + \beta)x(t)$. Since $1 + \beta > 1$, this will grow exponentially—the system is *unstable*. Anderson puts the system "in a box" by having hard bounds $[-1, +1]$ on the values of $x(i, t)$ in the original coordinates. As we see from

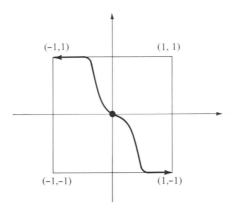

Figure 5.8   Phase-plane portrait of the BSB for the two-cell case.

Figure 5.8, the hard bounds destroy linearity. The trajectory heads for an edge, then follows it to a corner. This breaks the space into *attractors* or *capture regions*. Inputs are thus *classified* by the corners of the box, i.e., by amplifying the relative strengths of the original principal components. There are thus up to 2^n classifications in an n-element net—although, in general, not all corners will be stable attractors—and these classification can be read by the "next net" that receives the output of the recurrent matrix.

5.3 Hopfield Nets and Boltzmann Machines

In the one-layer perceptron model of Sections 4.2 and 4.4, the Hebb model of Section 4.3, and the associative nets/synaptic matrices of the previous section, the only adaptive elements have been evaluated directly. A major development of the 1980s has been the introduction of techniques that can also train "hidden units" (Figure 5.9) for which no training signal is directly available. In the rest of this chapter, we study three such models: Hinton and Sejnowski's Boltzmann machines; Barto's associative reinforcement learning; and Rumelhart, Hinton, and Williams' back-propagation algorithm.

As we have seen, a number of workers in AI are now adopting approaches designed for fault-tolerant computing in highly parallel networks. These networks are motivated by studies of neural networks, but are not designed to model the details of real brains. For example, we saw (Figure 5.1) that the connectionist solution to the problem of getting a 3-D interpretation from a 2-D sketch is to build a network such that, for each edge in the drawing, the interpretation network has a unit for each vertex interpretation, with excitatory connections for each 3-D edge compatible with them. Such a network thus represents plausible inferences by connection strengths. If designed

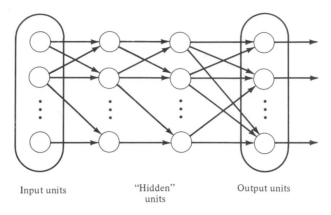

Input units   "Hidden"   Output units
                      units

Figure 5.9   A network with "hidden" units.

properly, only those nodes will eventually remain active that correspond to
the edges in a 3-D interpretation.

Given such a network, it is, of course, not obvious whether it will work
"as advertised." We must ask: Will the net converge to a state, and will that
state represent the best interpretation the image? How long does it take to
converge? More generally, we inquire: Where do the connection strengths
come from? In general, we need a learning algorithm. We devote this section
to the answer to these questions given by the specific theory developed by
Hopfield, Hinton, and Sejnowski. (See Geman and Geman, 1984, for a
rigorous analysis of the methodology and Marroguin, 1985, for extensive
demonstrations of its applicability in computer vision.) They use as unit the
familiar Binary Threshold Unit (McCulloch–Pitts neuron) whose output
is 1 iff $\sum w_{ij}s_i > \theta_j$ and is otherwise 0, where s_i is the current value of the ith
input and w_{ij} is the corresponding synaptic weight from i to unit j, whose
threshold is θ_j. In the McCulloch–Pitts nets of Chapter 2, every neuron processes its inputs to determine a new output at each time step. By contrast, a
Hopfield net is a net of such units subject to the *updating rule*: "Pick a unit at
random. If the sum of the weights on connections to other *active* units is
positive, turn it on. Otherwise turn it off." Moreover, Hopfield assumes
symmetric weights ($w_{ij} = w_{ji}$), as is true in most vision algorithms, where
constraints can be formulated in terms of symmetric weights plus threshold
terms.

Hopfield's contribution was to associate with such a net a measure called
the *Energy*. In the following definition, the first term is due to Hopfield, the
second to Hinton and Sejnowski:

$$E = -\frac{1}{2}\sum_{i \neq j} s_i s_j w_{ij} + \sum_i s_i \theta_i. \tag{6}$$

Note that when a given s_i goes from 0 to 1, the "energy gap," or change in E,
is given by $\Delta E = \sum w_{ij}s_j - \theta_i$.

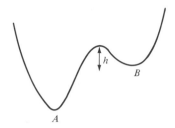

Figure 5.10   An energy landscape, in which energy increment h is required to escape from the valley of the local minimum B to enter the attactor of the global minimum A.

Thus Hopfield's updating rule is equivalent to the following: "Randomly pick a unit. Change its state (from 0 to 1, or vice versa) just in case doing so will lower the energy." Thus the state of the net changes until it enters a state of minimal energy in the sense that no change in any *one* of the variables s_i will lower the value E. We stress that there may be different such states—they are *local* minima. Global minimization is not guaranteed. For example, if we think of the curve in Figure 5.10 as an energy surface, the system might move to the local minimum B rather than the global minimum A.

Note that the above expression for ΔE crucially depended on the symmetry condition $w_{ij} = w_{ji}$, for without this condition the expression would instead be $\Delta E = \frac{1}{2} \sum (w_{ij} + w_{ji})s_j - \theta_i$ so that Hopfield's updating rule would not yield a passage to energy minimum in this case, but might instead yield a limit cycle, which (e.g. Kelso and Tuller, 1984) might be useful in modeling control of action (Figure 5.11). In a static pattern-recognition problem (as in Figure 5.1) a constraint w_{ij} will express the logarithm of the likelihood that units i and j are compatible in a coherent interpretation, so that $w_{ij} = w_{ji}$ is appropriate. In a control problem, however, a link w_{ij} might express the likelihood that the action represented by i should precede that represented by j, and thus $w_{ij} = w_{ji}$ is normally inappropriate.

To use the Hopfield net (symmetrical connections) to compute: Designate certain units as input, clamp their values so that the updating rule is not applied to them. Now repeatedly apply updating to all the other units to let the network settle to its local energy minimum; and then read the values of certain designated output units. Since there is no algorithm specifying which unit changes at the next time step, the corresponding input output function may yield different values depending on the sequence of choices made. The aim, given a search problem, is to so choose weights for the network that E is a measure of the overall constraint violation. We can then use cooperative computation to maximize constraint satisfaction. But, as we have already noted, the net only gives *local* minima.

To see the solution offered by Hinton, Sejnowski, and Ackley, 1984 consider how one might get a ball-bearing traveling along the curve in Figure 5.10 to *probably* end up in the deepest minimum. The idea is to shake the box

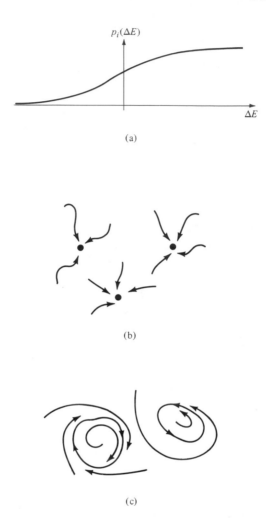

Figure 5.11   (a) Logistic function p used for the Boltzmann machine. (b) Typical flow map of neural dynamics for symmetric connections ($w_{ij} = w_{ji}$). (c) More complicated dynamics that can occur for unrestricted w_{ij}. Limit cycles are possible. ((b) and (c) adapted from Hopfield and Tank, 1986).

"about h hard"—then the ball is more likely to go from B to A than from A to B. So, on average, the ball should end up in A's valley. Kirkpatrick et al. (1983) (after Metropolis et al., 1953) developed this into a general method called "simulated annealing" for making likely the escape from local minima by allowing jumps to higher energy states. In the statistical theory of gases, the gas is described not by a deterministic dynamics, but rather by the probability that it will be in different states. The 19th century physicist Ludwig Boltzmann developed a theory that included a probability distribution for the states of a gas when it had reached a uniform distribution of temperature (i.e., every small region of the gas had the same kinetic energy). The rather daring claim of Hinton and Sejnowski [see Cragg and Temperley (1954) for a precursor of this idea] was that this distribution might also be used to describe neural interactions, where low temperature T is replaced by a small noise term T (the neural analog of random thermal motion of molecules).

At thermal equilibrium at temperature T, the Boltzmann distribution gives as the relative probability that the system will occupy state A, as against state B,

$$\frac{p_A}{p_B} = e^{-(E_A - E_B)/T} = \frac{e^{E_B/T}}{e^{E_A/T}} > 1 \quad \text{if } E_B > E_A, \tag{7}$$

i.e., the "better" the minimum, the more likely the system is to occupy it. Since thermal equilibrium is defined as the situation in which the *probability distribution* is stable, there is *no* claim that the system has reached a single stable state. Hinton and Sejnowski applied simulated annealing to Hopfield nets by modifying the rule for state transition as follows:

Pick a unit at random. Compute $\Delta E = \sum s_j w_{ij} - \theta_i$. Set s_i to 1 with probability $p_i(\Delta E) = 1/(1 + e^{-\Delta E/T})$, as shown in Figure 5.11.

Note that this converges to the "old" rule, a step function with the step at $\Delta E = 0$, as $T \to 0$.

The approach to thermal equilibrium for high T is fast, but p_A/p_B is small; for low T, thermal equilibrium is reached slowly, but p_A/p_B is large. Noise can make the system perform better! These nets work well for constraint-satisfaction searches; can be implemented very naturally in a parallel network; and give "analog" Bayesian inference, as shown in Hinton and Sejnowski, 1983. They are examples of what statisticians call Markov random fields (cf. Geman and Geman, 1984).

Learning

There are two extremes to the study of learning: the classic approach, exemplified by the perceptron, is to posit that the system *self-organizes* on the basis of a single uniform learning rule, no matter what the task. At the other extreme, many practicioners in AI leave little role for learning, requiring much "knowledge" be coded in a domain-specific database. Hinton and Sejnowski offer an approach at the first extreme.

The learning procedure: Given a net with designated input and output units, the following procedure will enable the net to learn (a good approximation to) a designated conditional probability distribution on the set of (input vector, output vector) pairs:

Positive Phase
(a) Clamp the input vector and output vector. In other words, as units are updated at random, these values are not allowed to change.

(b) Let the net reach thermal equilibrium.
(c) *Increment* the weight between two units every time they are *both* on together.
   Negative Phase. This phase is added as one approach to stopping the the weights from saturating, since the above rule only increases them.
(a) Clamp the input values but *not* the output.
(b) Let the net settle.
(c) *Decrement* the weight between two units every time they are both on together.

The positive phase is the familiar Hebb rule. Here, it is the negative phase (suggested by Francis Crick) that provides the necessary antidote to saturation of the synaptic weights (cf. Section 4.3).

The required behavior is a set of conditional probabilities $\{P^1(O_\beta/I_\alpha)\}$, where I_α is an input vector and O_β is an output vector. Given a particular set of weights, the network will exhibit a particular set of conditional probabilities when the input is clamped, namely, $\{P^0(O_\beta/I_\alpha)\}$. The learning procedure is guaranteed to make the $P^0(O_\beta/I_\alpha)$ for the resultant weights more like the $P^1(O_\beta/I_\alpha)$. Using the information measure,

$$G = \sum_{\alpha, \beta} P^1(O_\beta/I_\alpha)\ln\frac{P^1(O_\beta/I_\alpha)}{P^0(O_\beta/I_\alpha)} \tag{8}$$

we change the weights by

$$\Delta w_{ij} = -\varepsilon\frac{\partial G}{\partial w_{ij}}\frac{\varepsilon}{T}(\langle_1 s_i s_j\rangle - \langle_0 s_i s_j\rangle) \tag{9}$$

where $\langle_1 s_i s_j\rangle$ is the expected value of $s_i s_j$ during the positive phase, and $\langle_0 s_i s_j\rangle$ is the expected value during the negative phase. The first term is like a Hebb synapse, the second is "unlearning" and occurs when the network is "hallucinating" the output.

In fact, $\partial G/\partial w_{ij}$ only depends on the P_{ij}'s for the given i and j. This result is like an assignment of credit—the change of weight at one place may depend on remote weights, yet we still get the $\partial G/\partial W$ result. So, we can predict how a change in a "microscopic" variable w_{ij} affects the "macroscopic" probabilities P^0. This remarkable simplification occurs because, at thermal equilibrium, each *weight* contributes *independently* to the probability ratio of global configurations.

In summary, the Boltzmann machine interleaves two searches:

Inner loop: Anneal to reach equilibrium at a *finite* temperature (with some units clamped). This finds states that have low energy, but does not stay in any one state.

Outer loop: Use the statistics measured at equilibrium with many different clamped inputs to compute a single move in the search for a good set of weights (i.e., a good energy landscape for this task). Repeated iterations of the outer loop lead to low values of G (but not the minimum, when, as in general, we cannot guarantee that G is convex).

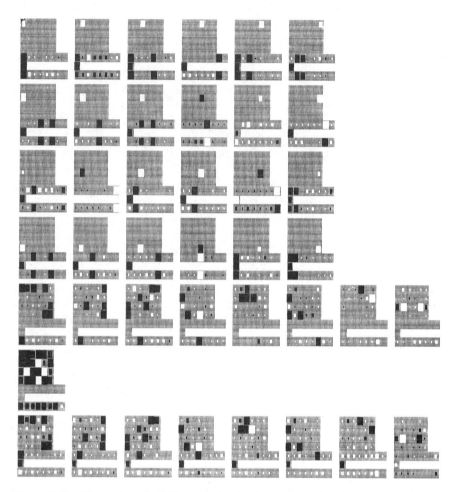

Figure 5.12 Network with 41 binary units that learned to solve the negation problem using the Boltzmann machine learning procedure. For details, see text. (Courtesy of T. Sejnowski.)

Figure 5.12 shows a net that uses the Boltzmann machine learning procedure to solve the negation problem. The problem is to provide an output that is the negation of the input if the negation bit is set to one, otherwise the output should be identical to the input. (Note that this is equivalent to performing eight simultaneous exclusive ORs.) The diagram shows the weights for a network with 8 input units (bottom row), 8 output units (third row from bottom), 1 negation bit (second row from bottom), and 24 hidden units (top four rows) that had connections to all the other units, including a *true* unit whose value was always 1. The weights are shown as small squares within each unit. The color indicates the sign (white for a positive or excitatory connection, black for a negative or inhibitory connection), and the area of the square is proportional to the magnitude of the weight. Note that the shape of each unit resembles the overall shape of the entire diagram. The position of the weight within the unit is the relative position of the unit it is connected to. The weight to the true unit is located in the position within the unit that the unit itself occupies in the diagram. The largest weights are about 100 in magnitude. The network was trained using the Boltzmann learning procedure described in the text with the temperature fixed at 10. For each presentation, the input units and sign unit were turned on randomly with a probability of 0.5. After training on 15,000 randomly generated patterns updating the weights after every 15 patterns, the network produced the correct answer for all output units 98% of the time using a slow annealing schedule. This is an example of a problem that is not linearly separable and cannot be solved with a perceptron having only one layer of variable weights; it is a second-order predicate in the sense of Minsky and Papert, 1969. Note, however, that the input units are independent of each other, and that most of the hidden units discovered codes that are specialized for one or at most two input-output pairs.

5.4 Reinforcement Learning and Back-Propagation

Barto and colleagues have developed networks that employ reinforcement learning instead of supervised learning (Barto, Sutton, and Brouwer, 1981; Barto, Anandan, and Anderson, 1985; Barto and Anandan, 1985; Barto, 1985). Where a unit that learns by a supervised learning rule requires a teacher to tell it what its actions should be, a unit that learns by a reinforcement learning rule requires a "critic" that can criticize the action it has taken, providing, for example, "reward" for a good action and "penalty" for a bad one. As Barto and colleagues have shown, if a unit can learn to increase the frequency of reward from a very "noisy" critic, it can act cooperatively with other units in a network to improve the performance of the entire network.

Barto's reinforcement learning method is based on the A_{R-P}, associative reward–penalty unit (Barto and Anandan, 1985), shown in Figure 5.13(a). Here, r takes value $+1$ for "reward" and -1 for "penalty." The value $+1$ thus means "do it like you just did it," while for the perceptron, a training input value of $+1$ means "y should be 1." The $x_{j(t)}$ may be any real numbers; whereas the action/output y is binary, $+1$ or -1. We compute $s(t)$ by the linear summation $s(t) = \sum w_j x_j(t)$, and then form $y(t)$ on the basis of a random number $\mu(t)$, which is construed as a random threshold or as noise in the membrane potential, by the rule

$$y(t) = \begin{cases} +1 & \text{if } s(t) + \eta(t) > 0; \\ -1 & \text{else.} \end{cases}$$

The weights are changed according to the rule

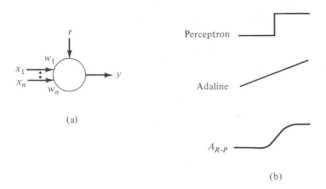

(a)

(b)

Figure 5.13 (a) An A_{R-P} element with reinforcement signal r. (b) The effective reinforcement curves in three different learning schemes.

$$\Delta w_i(t) = \begin{cases} \rho[y(t) - E[y(t)|s(t)]]x_i(t) & \text{if } r(t) = +1, \\ \rho\lambda[-y(t) - E[y(t)|s(t)]]x_i(t) & \text{if } r(t) = -1, \end{cases} \tag{10}$$

where $\rho > 0$ and $0 < \lambda < 1$.

The expectation term in this rule depends on the form of the noise $\eta(t)$. Uniformly distributed noise gives the expectation term a linear form that makes the rule similar to the Adaline [Figure 5.13(b)], except that it would saturate at values corresponding to the bounds of the interval from which the noise value could be drawn. If the noise $\eta(t)$ is always 0, in other words, if there is no noise, then $y(t) = E[y(t)|s(t)]$, and so we have

$$\Delta w_i(t) = \begin{cases} 0 & \text{if } r(t) = +1, \\ -2\lambda\rho y(t)x_i(t) & \text{if } r = -1, \end{cases} \tag{11}$$

and this is just the perceptron rule. If the noise is drawn from a normal, or similar, distribution, then the expectation term has a sigmoid form shown in Figure 5.13b.

For $r(t) = +1$, the rule gives what Widrow, Gupta, and Maitra (1974) call "positive bootstrapping," which applies the learning term that would have been invloved had the response been correct. This decreases the discrepancy between the action $y(t)$ and the expected action. For $r(t) = -1$, the system is adjusted toward $-y(t)$, the action it did *not* do ("negative bootstrapping"). Setting $\lambda < 1$ decreases the size of the change when there is a penalty.

Consequently, the reinforcement signal switches the unit between positive and negative bootstrapping mode, a process Widrow et al. (1974) called "selective bootstrap adaptation." In the case of an A_{R-P} unit, however, the weight changes alter the unit's action probabilities, with reward making the unit more likely, in response to similar input patterns, to do what it just did, and penalty making this action less likely.

The probabilistic nature of an A_{R-P} unit allows it to maximize reward frequency when the critic itself is very noisy. Barto and Anandan (1985) define the following *associative reinforcement learning task*. Given $d(x, y)$ as the probability of reward for input x and output y, the task is to adjust the weights so that input x will yield the output $y(x)$ for which $d(x, y(x)) \geq d(x, y)$ for all outputs y, i.e., such that $y(x)$ maximizes the probability of reward. Note that the maximum reward probability $d(x, y(x))$ may still be less than 0.5—just consider tossing coins all of which are biased to come up tails, although heads are rewarded. Both $y = +1$ and $y = -1$ can have a nonzero probability of reward, and determining which action is best gets difficult if these probabilities are close. Barto and Anandan (1985) proves the following theorem:

Convergence Theorem. *If the input patterns are linearly independent, the noise distribution function is continuous and strictly monotonic, each pattern has nonzero probability, and the step size ρ_t decreases with time t such that $\sum \rho_t = \infty$, $\sum \rho_t^2 < \infty$, with $0 < \lambda \leq 1$ fixed, then the system will, with probability 1, pick the better action with probability > 0.5 for each x. As $\lambda \to 0$, the asymptotic probability of doing the better action will go to 1.*

The parameter λ is somewhat analogous to the T of simulated annealing in that it controls the equilibrium action probabilities.

Barto's current studies address networks of A_{R-P} elements, handling the hidden unit problem. The net of Figure 5.14(a) is the simplest example showing how A_{R-P} units exhibit a kind of cooperativity when reinforcement is broadcast to all units. In this example, the critic broadcasts reward whenever the network acts as the identity function. To each of the units, the problem appears to be one of learning in a very noisy environment, the noise being introduced by the activity of the *other* unit. However, since A_{R-P} units can learn in the presence of a very noisy critic, the units in this network discover how to control reinforcement by linking up appropriately, i.e., by realizing the desired identity function. Another problem was to train the net of Figure 5.14(b) to become an XOR. This succeeded, too, for $\lambda \neq 0$. One way in which the network solves this problem is that u_2 learns the linearly separable function $x_1 \vee x_2$, u_1 learns the "wrong" case $x_1 = x_2 = 1$, and then u_2 adapts to this correction term.

Of those surveyed here, Barto's scheme may be the most realistic in terms of neuroscience. Williams (1986) has shown that for arbitrary nonrecurrent

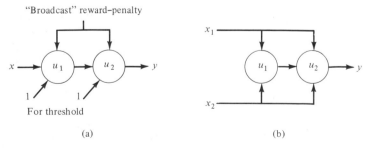

Figure 5.14   Two nets exhibiting reinforcement training of hidden units to yield (a) the identity function, (b) an XOR gate.

networks of A_{R-P} units with $\lambda = 0$, each weight changes according to an unbiased estimate of the gradient of the probability of overall network reward with respect to the weight. Hidden units thus achieve gradient-following without the nonphysiological back-propagation of the scheme to be described below. However, it turns out that with $\lambda = 0$, the case covered by Williams' result, networks tend to get stuck at local maxima of the reward probability. Making λ a small positive value seems to cure this difficulty. The parameter λ may control the search in weight space in much the same manner that computational temperature T controls the search in activation pattern space in the case of the Boltzmann machine, but additional theory needs to be developed to adequately resolve this issue.

Another study demonstrating reinforcement learning addresses the problem of learning when no teacher is available to evaluate every action. A classic example (Samuel, 1959) is playing checkers: How does one evaluate a move in the middle of the game, when the only evaluation ("win" or "lose") comes at the end? This is known as the "assignment of credit problem"—How do we assign credit to individual moves, when only the total pattern of play is evaluated? One approach reminds me of a scene from a Marx Brothers' movie in which Groucho said "Go search the house next door"; his brothers replied "There is no house next door"; to which Groucho replied "Build a house, and then search it!"

The same solution is offered here: "Climb a hill" (i.e., maximize some payoff function such as $-E$ for some energy measure E). "There is no hill." "Build a hill, and then climb it!"

Barto, Sutton, and Anderson (1982) demonstrated that connectionist methods can handle credit assignment in this way by applying adaptive units to the task of balancing the simulated pole shown in Figure 5.15 by applying

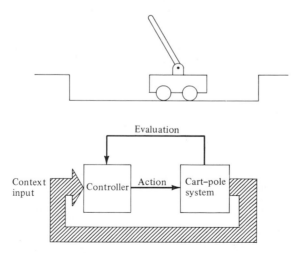

Figure 5.15   An example raising the problem of delayed reinforcement: the inverted pendulum (pole-balancing).

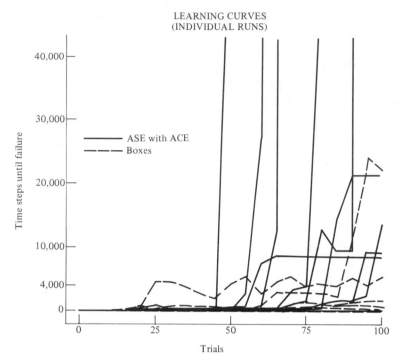

Figure 5.16   Learning curves for reinforcement learning and the pole-balancing problem.

forces to the cart. Figure 5.15 also shows a controller made of two different types of adaptive elements that each receive as input a coding of the state of the cart–pole. The "associate search element" (ASE), works very much like an A_{R-P} unit, and the "adaptive critic element" (ACE) receives a failure signal when the pole falls or the cart hits the end of the track. The ACE constructs a more informative "internal reinforcement signal" that makes the ACE's learning task much easier. It is this signal that is the hill the system builds. Figure 5.16 shows how the system's performance improves with experience by showing how the length of time until failure increases as a function of trials, where a trial is a period of balancing that ends with failure. The other curves in the figure show the performance of several other systems with which the ASE/ACE controller was compared:

$$\text{actions} = \pm \text{ force on cart,}$$

$$\text{context} = \text{state of cart–pole} = (x, \dot{x}, \theta, \dot{\theta}),$$

where x is the position of the cart and $\dot{x}$ is the corresponding velocity; θ is the angle of the pole and $\dot{\theta}$ is the corresponding angular velocity.

The task is to apply forces to move the cart in such a way that it never hits the end stops, and the pole never falls over. Thus, evaluation = failure when the pole falls or the cart hits the end of track; otherwise no evaluative feedback.

The back-propagation algorithm

Rumelhart, Hinton, and Williams (1983) address the problem of hidden units as follows. Each unit has both input and output taking continuous values in some range $[a, b]$. The response is a sigmoidal function of the weighted sum. Thus, if a unit has inputs x_k with corresponding weights w_{ik}, the output y_i is given by

$$y_i = f(\sum w_{ik} x_k) \quad \text{where } f \text{ is the sigmoid } f(x) = \frac{1}{(1 + e^{-x})}. \quad (12)$$

The units are formed into a net with only certain units being "visible," i.e., supplying output from the net. The environment only directly affects the visible units. We are given a set $\{p\}$ of input patterns, and for each p a corresponding desired target pattern t^p for the output units. With o^p the actual output pattern elicited by input p, the aim is to adjust the weights in the networks to minimize the error

$$E = \sum_{\text{patterns } p} \sum_{\text{output neurons } i} (t_i^P - o_i^P)^2. \quad (13)$$

Rumelhart et al. devised a formula for propagating back the gradient of this evaluation from a unit to its inputs, and this can continue by back-propagation through the entire net (Figure 5.17). Note that the error measure, E, used in back-propagation is similar to the performance measure, G, used in Boltzmann machines and has no relation to the energy, E, used in Boltzmann machines.

Consider a net with the neurons arranged in $m + 1$ layers, with $m + 1$ the

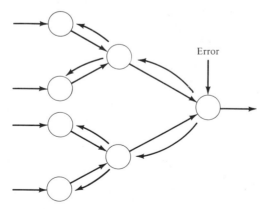

Figure 5.17   The back-propagation method of Rumelhart, Hinton, and Williams.

output layer and m the last hidden layer. The method uses gradient descent:

$$\Delta w_{ij}^{(m)} \sim - \frac{\partial E}{\partial w_{ij}^{(m)}} = k \cdot d_j^{(m+1)} p_i^{(m)} \tag{14}$$

where $d_j^{(m+1)}$ is an error measure from level $m+1$. The method then propagates back. At the output, this is a generalization of the Widrow–Hoff rule, where Widrow–Hoff use $[y - \sum w_i x_i] x_i$, Rumelhart et al. use $f[y - \sum w_i x_i] y (1 - y) x_i$. The problem is to show how to do this by local computations, while avoiding local minima. The output $d^M = (t_j - p_j) f_j^{(m)}$ $\cdot (net_j^m)$ is the product of the error, the derivative of the activation, and the summed input, respectively. For an internal unit, the d's are obtained recursively

$$d_j^{(n)} = [\sum d_k^{(n+1)} w_{jk}^{(n)}] f_j^{(n)} (net_j^n). \tag{15}$$

Although no formal proof is available, simulation shows that the scheme avoids many of the false minima that bedeviled other methods. Rumelhart et al. found the algorithm to be rather robust when tested on several hundred problems. In a few cases, local minima were found, but they were easy to escape from. If there are enough units, the method can attain $E = 0$; if not, the weights can be trained to minimize E, a result that is optimal in the sense of least squares.

References for Chapter 5

Amari, S., 1977a, A mathematical approach to Neural Systems, in *Systems Neuroscience* (J. Metzler, Ed.), Academic Press.

Amari, S., 1977b, Neural theory of association and concept-formation, *Biol. Cybern.* **26**: 175–185.

Anderson, J.A., 1983, Cognitive and psychological computation with neural models, *IEEE Transactions Systems, Man and Cybernetics* **13**: 799–815.

Arbib, M.A., 1964, *Brains, Machines and Mathematics*, McGraw-Hill.

Arbib, M.A. and Caplan, D., 1979, Neurolinguistics must be computational, *Behavioral and Brain Sciences* **2**: 449–483.

Ashby, W.R., 1954, *Design for a Brain*, Chapman and Hall.

Ballard, D.H., 1986, Cortical connections and parallel processing: Structure and function, *Behavioral and Brain Sciences* **9**: 67–120.

Ballard, D.H., Hinton, G.E., and Sejnowski, T.J., 1983, Parallel visual computation, *Nature* **306**: 21–26.

Ballard, D.H. and Brown, C.M., 1982, *Computer Vision*, Prentice-Hall.

Barto, A.G., 1985, Learning by statistical cooperation of self-interested neuron-like computing elements, *Human Neurobiology* **4**: 229–256.

Barto, A.G., and Anandan, P., 1985, Pattern recognizing stochastic learning automata, *IEEE Trans. Systems, Man and Cybernetics* **15**: 360–375.

Barto, A., Anandan, P., and Anderson, C., 1985, Cooperativity in networks of pattern recognizing stochastic learning automata, *Proceedings of the 4th Yale Workshop on Applications of Adaptive Systems Theory*, New Haven, CT, May.

Barto, A., Sutton, R., and Anderson, C., 1982, Spatial learning simulation systems, *Proceedings of the 10th IMACS World Congress on Systems Simulation and Scientific Computation*, pp. 204–206.

Barto, A.G., Sutton, R.S., Brouwer, P., 1981, Associative search network: A reinforcement learning associative memory, *Biol. Cybern.* **40**: 201–211.

Cannon, W.B., 1939, *The Wisdom of the Body*, Norton.

Cottrell, G., and Small, S., 1983, A connectionist scheme for modeling word sense disambiguation, *Cognitive Science* **6**: 89–120.

Cragg, B.G., and Temperley, H.N.V., 1954, The Organization of neurones: A cooperative analogy, *EEG Clin. Neurophysiol.* **6**: 85–92.

Erman, L.D., Hayes-Roth, F., Lesser, V.R., and Reddy, D.R., 1980, The HEARSAY-II speech understanding system: Integrating knowledge to resolve uncertainty, *Computing Surveys* **12**: 213–253.

Feldman, J.A., 1981, A connectionist model of visual memory, in *Parallel Models of Associative Memory* (G.E. Hinton and J.A. Anderson, Eds.), Lawrence Erlbaum Associates.

Feldman, J.A., and Ballard, D.H., 1982, Connectionist models and their properties, *Cognitive Science* **6**: 205–254.

Geman, S., and Geman, D., 1984, Stochastic relaxation, Gibbs distributions, and the Bayesian restoration of images, *IEEE Trans. on Pattern Analysis and Machine Intelligence* **6**: 721–741.

Gigley H.M., 1983, HOPE-AI and the dynamic process of language behavior, *Cognition and Brain Theory* **6**: 39–88.

Greene, P.H., 1969, Seeking mathematical models of skilled actions, in *Biomechanics* (H.C. Muffley and D. Bootzin Eds.), Plenum, pp. 149–180.

Grossberg, S., 1970, Neural pattern discrimination, *J. Theor. Biol.* **27**: 291–337.

Grossberg, S., 1978, Communication, memory, and development, in *Progress in Theoretical Biology* (R. Rosen and F. Snell, Eds.), Academic Press, Vol. 5, pp. 183–232.

Grossberg, S. and Levine, D., 1975, Some developmental and attentional biases in the contrast enhancement and short term memory of recurrent neural networks, *J. Theor. Biol.* **53**: 341–380.

Haken, H., 1978, *Synergetics: An Introduction. Nonequilibrium Phase Transitions and Self-Organization in Physics, Chemistry, and Biology* (Second Edition), Springer-Verlag.

Hinton, G.E., and Anderson, J.A., Eds., 1981, *Parallel Models of Associative Memory*, Lawrence Erlbaum Associates.

Hinton, G.E., Sejnowski, T.J., and Ackley, D.H., 1984, A learning Boltzman machine, *Cognitive Science.* **9**: 147–169.

Hinton, G.E., and Sejnowski, T.J., 1983, Optimal perceptual inference, in *Proceedings of the IEEE Computer Society Conference on Computer Vision and Pattern Recognition*, Washington, D.C., pp. 448–453.

Hopfield, J., 1982, Neural networks and physical systems with emergent collective computational properties, *Proc. Nat. Acad. Sci., USA* **79**: 2554–2558.

Hopfield, J.J. and Tank, D.W., 1986, Computing with neural circuits: A model, *Science* **233**: 625–632.

Hummel, R.A., and Zucker, S.W., 1983, On the foundations of relaxation labelling processes, *IEEE Trans. Pattern Analysis and Machine Intelligence* **5**: 267–287.

Kelso, J.A.S., and Tuller, B., 1984, A dynamical basis for action systems, in *Handbook of Cognitive Neuroscience* (M.S. Gazzaniga, Ed.), pp. 321–356.

Kilmer, W.L., McCulloch, W.S., and Blum, J., 1969, A model of the vertebrate central command system, *Int. J. Man-Machine Studies* **1**: 279–309.

Kirkpatrick, S., Gelatt, C.D., Jr., and Vecchi, M.P., 1983, Optimization by simulated annealing, *Science* **220**: 671–680.

Kohonen, T., 1977, *Associative Memory: A System-Theoretical Approach*, Springer-Verlag.

Kohonen, T., 1982, Self-organized formation of topologically correct feature maps, *Biol. Cybern.* **43**: 59–69.

Kohonen, T., 1984, *Self-Organization and Associative Memory*, Springer-Verlag.

Kohonen, T., and Oja, E., 1976, Fast adaptive formation of orthogonalizing filters and associative memory in recurrent networks of neuron-like elements, *Biol. Cybern.* **21**: 85–95.

Marroquin, J.L., 1985, *Probabilistic Solution of Inverse Problems*, AI-TR 860, MIT AI Lab.

McClelland, J.L., and Rumelhart, D.E., 1981, An interactive activation model of context effects in letter perception: Part 1. An account of basic findings, *Psych. Rev.* **88**: 375–407.

Metropolis, N., Rosenbluth, A.W., Rosenbluth, M.N., Teller, A.H., and Teller, E., 1953, Equation of state calculations for fast computing machine, *J. Chem. Phys.* **6**: 1087.

Rosenfeld, A., Hummel, R.A., and Zucker, S.W., 1976, Scene labelling by relaxation operations, *IEEE Trans. Systems, Man and Cybernetics* **6**: 420–433.

Rumelhart, D.E., Hinton, G.E., and Williams, R.J., 1986, Learning internal representations by error propagation, in Rumelhart and McClelland, 1986a, Vol. 1, pp. 318–362.

Rumelhart, D.E., and McClelland, J., 1982, An interactive activation model of context effects in letter perception, Part 2. The contextual enhancement effect and some tests and extensions of the model, *Psych. Rev.* **89**: 60–94.

Rumelhart, D., and McClelland, J., Eds., 1986a, *Parallel Distributed Processing: Explorations in the Microstructure of Cognition*, The MIT Press/Bradford Books.

Rumelhart, D.E., and McClelland, J.L., 1986b, On learning the past tense of English verbs, in Rumelhart and McClelland, 1986a, Vol. 2, pp. 216–271.

Samuel, A.L., 1959, Some studies in machine learning using the game of checkers, *IBM J. Res. and Dev.* **3**: 210–229.

Sejnowski, T.J., 1981, Skeleton filters in the brain, in *Parallel Models of Associative Memory* (G.E. Hinton and J.A. Anderson, Eds.), Lawerence Erlbaum Associates.

Selfridge, O.G., 1959, Pandemonium: A paradigm for learning, in *Mechanization of Thought Processes*, Her Majesty's Stationery Office, pp. 511–531.

Trehub, A., 1979, Associative sequential recall in a synaptic matrix, *J. Theor. Biol.* **81**: 569–576.

Trehub, A., 1987, in *Vision, Brain and Cooperative Computation* (M.A. Arbib and A.R. Hanson, Eds.), The MIT Press.

Waddington, C.H., 1965, *The Strategy of the Genes*.

Waltz, D.L., 1975, Understanding line drawings of scenes and shadows, in *The Psychology of Computer Vision* (P.H. Winston, Ed.), McGraw-Hill.

Waltz, D.L., and Pollack, J.B., 1985, Massively parallel parsing, *Cognitive Science* **9**: 51–74.

Widrow, B., Gupta, N.K., and Maitra, S., 1973, Punish/reward: Learning with a critic in adaptive threshold systems, *IEEE Trans. Systems, Man and Cybernetics* **5**: 455–465.

Widrow, G., and Hoff, M.E., 1960, Adaptive Switching Circuits, in *1960 IRE WESCON Convention Record*, Part 4, pp. 96–104.

Wiener, N., 1948, *Cybernetics: or Control and Communication in the Animal and the Machine*, The Technology Press and Wiley (Second Edition, The MIT Press, 1961).

Williams, R.J., 1986, *Reinforcement Learning in Connectionist Networks: A Mathematical Analysis*, Technical Report TR86–05, Institute for Cognitive Science, UCSD.

CHAPTER 6

Turing Machines and Effective Computations

6.1 Manipulating Strings of Symbols

In Chapter 2, we introduced the notion of a *finite automaton* by abstraction from the concept of a network of (McCulloch–Pitts) neurons. In this chapter, we wish to continue our study of automata in a more general setting. The *Oxford English Dictionary* defines an *automaton* (plural, *automata*) as "Something which has the power of spontaneous movement or self-motion; a piece of mechanism having its motive power so concealed that it appears to move spontaneously; now usually applied to figures which simulate the actions of living beings, as clockwork mice, etc." Today the computer has replaced the clockwork mouse as the archetype of the automaton; and with it, our emphasis shifts from simulation of motion to simulation of information processing, although this will change again with the increasing importance of robotics. Automata theory, in its widest sense, might now embrace such diverse activities as the building of a space station's control system or the programming of a computer to play chess. In the theory of *abstract* automata, we are less concerned with the design of automata to do specific tasks, and more concerned with understanding the capabilities and limitations of whole *classes* of automata. Our aim here is to develop enough abstract automata theory to allow us to answer interesting questions about the relationships among brains, machines, and mathematics.

Automata theory has already provided the following:

1. A characterization of *all* computable function (e.g., as those computable by Turing machines)—it being now a highly active area of research to find which of the concomitant computations are *practical*—together with the

demonstration that no computer can compute, of an arbitrary computer, whether or not that second computer will ever halt.

2. The demonstration of *universality*—that there is a computer which can do the job of any other computer provided that it is suitably programmed.
3. Parsing systems of formal languages, and concomitant automata, which form the basis for a rigorous treatment of compilers for computer languages.

We shall study the first two issues in this chapter [a far fuller account is given by Kfoury, Moll, and Arbib (1982)]; for an account of the formal theory of both programming languages and natural languages, see Moll, Arbib, and Kfoury (1987).

Consider the item displayed on the next line:

$$1\ 0\ 0\ 1\ 1$$

it is a string (we use "string" as a synonym for "sequence") of five symbols, of which the first, fourth, and fifth are 1's, while the second and third are 0's. Whether we choose to interpret this string of symbols as a decimal number or binary number, or as something else, depends on our "mental set," on the context. To a binary computer 1 0 0 1 1 is "nineteen"; to a decimal computer it is "ten thousand and eleven." Similarly, the function which places 0 at the end of a string of 0's and 1's is, to a binary computer, "multiplication by two," whereas to a decimal computer, it is "multiplication by ten."

The point I am trying to make, then, is the familiar one that computers are symbol-manipulation devices. What needs further emphasis is that they can thus be numerical processors, but *the numerical processing that they undertake is only specified when we state how numbers are to be* **encoded** *as strings of symbols, which may be fed into the computer, and how the strings of symbols printed out by the computer are to be* **decoded** *to yield the numerical result of the computation.*

Our emphasis in what follows, then, is on the ways in which information-processing structures (henceforth called automata) transform strings of symbols into other strings of symbols. Sometimes it will be convenient to emphasize the interpretation of these strings as encodings of numbers, but in many cases we shall deem it better not to do so.

As before, we shall usually use X to denote the *input alphabet*, the set of symbols from which we may build up strings suitable for feeding into our automaton. The symbol Y will usually denote the *output alphabet*, our automaton emitting strings of symbols from Y. Moreover, given any set A, A^* denotes the set of all finite sequences of elements from A, and we'll call the number of symbols, $|\alpha|$, in a sequence α the *length* of α. For mathematical convenience, we shall include in A^* the *empty sequence* Λ of length 0. We need Λ for the same reason that we had to invent the number 0. Just as it became distinctly unhelpful to write "x with nothing added to it" instead of "$x + 0$" so we prefer to say "input Λ" rather than "no input was supplied." It allows us to state many theorems in general form, without having to treat "no input" as a special case. Given two sequence $\alpha = a_1 \cdots a_n$ and $\beta = b_1 \cdots b_m$ we may *concatenate* them to obtain $\alpha \cdot \beta = a_1 \cdots a_n b_1 \cdots b_m$, and for all α we set $\alpha \cdot \Lambda = \Lambda \cdot \alpha = \alpha$.

Thus X^* will usually denote the set of all input strings to our automaton, and Y^* will indicate a set that includes all possible output strings of our automaton.

Our general notion of an automaton, then, is a device to which we may present a string of symbols from X, i.e., an element of X^*. If and when the machine finishes computing on this string, the result will be an element of Y^*. We say "if" because certain input strings may drive the computer into a "runaway" condition—e.g., endless cycling through a loop—from which a halt is impossible without external intervention (which amounts to changing the input string). This case might correspond to associating with the machine a device that produces an infinite string of elements of Y for each string in X^*— but, in fact we shall not treat this viewpoint further in this chapter.

Thus, in a very general form, we may say that automata theory is the study of *partial* functions $F: X^* \to Y^*$—that is, ways whereby *some* of the strings in X^* have assigned to them output strings from Y^*, it being understood that for other input strings w, the function $F(w)$ may not be defined at all. However, such a function becomes truly a part of "classical" automata theory only if we can relate it to a *finitely specifiable substrate*—or if we are eager to prove that no such substrate exists for it. Of course, once one has developed a body of theorems, one sees how they can be generalized if the finiteness condition is removed, so this criterion does not cover all of present-day automata theory.

Before we specify in more detail the forms (of which the Turing machine, to be introduced in Section 6.2, is one) of substrate that have figured most prominently in automata theory, it is useful to distinguish *on-line* machines from *off-line* machines. An on-line machine is one that may be thought of as processing data in an interactive situation—in processing a string it must yield a continual flow of outputs, processing each symbol completely (albeit in a way dependent on prior inputs) before it reads in the next symbol. This means that the corresponding function $F: X^* \to Y^*$ must have the following special property:

1 Definition. F is a *sequential function* if, for each nonempty string u of X^*, there exists a function $F_u: X^* \to Y^*$ such that for every nonempty v in X^*

$$F(uv) = F(u) \cdot F_u(v);$$

that is, the input string u causes the machine to put out the string $F(u)$ and to "change state" in such a way that it henceforth processes inputs according to a function F_u determined solely by F and u.

If we define $f(\Lambda) = F(\Lambda)$ whereas, for $x \neq \Lambda$, the function $f(x)$ is the substring of $F(x)$ produced in response to the last symbol of x, we see that $f: X^* \to Y^*$ allows us to reconstruct F by the formula

$$F(x_1 x_2 \cdots x_n) = f(\Lambda) f(x_1) f(x_1 x_2) \cdots f(x_1 x_2 \cdots x_n)$$

if each $x_1, \ldots, x_n$ is in X. Conversely,

$$f(x_1 x_2 \cdots x_n) = F_{x_1 x_2 \cdots x_{n-1}}(x_n).$$

Let us define, for any u in X^*, the function $L_u: X^* \to X^*$, which simply places u to the left of any string: $L_u(x) = ux$. Then we see that f_u, the function corresponding to F_u, has the simple form:

$$f_u(x) = fL_u(x),$$

and for a string uv we find, by changing f to f_u and u to v above, that

2                          $$f_{uv}(x) = f_u L_v(x) = fL_u L_v(x).$$

It thus makes sense to speak of each function f_u for u in X^* as a *state* of the sequential function F—so that $f = F_\Lambda$ is the *initial state* of the sequential function F—with the input v serving to change state f_u to state $f_u L_v$ while the output corresponding to state f_u is $f_u \Lambda = fL_u(\Lambda) = f(u)\Lambda_v$. Thus the evaluation of F may be captured by the input-state–output-sequence shown in Figure 6.1.

Three portraits of a sequential machine are shown in Figure 6.2. Since we are in the habit of considering the first letter of a string to be the leftmost letter (translators of this text into Hebrew, please take care!), it seems appropriate (although perhaps a majority of authors use the opposite convention) to have the input line drawn to the right of the box—so that the first input symbol is the first to reach the box—and the output line drawn to the left of the box—so that the last output symbol is the last to leave the box.

Given F, let Q_F be the set $\{g \mid g = fL_u \text{ for some } u \in X^*\}$, the set of distinct states of F. We can then define a next-state function $\delta(f_u, X) = f_u L_x$ and an output function $\beta(f_u) = f_u(\Lambda)$. (This type of construction should be familiar from the discussion of machine realization in Section 3.2.) Thus a sequential function yields a special case of a sequential machine, where: A *sequential machine M* is specified by three sets, the set X of inputs, the set Y of outputs, and the set Q of states together with a next-state function $\delta: Q \times X \to Q$ and an output function $\beta: Q \to Y$. If we have $\beta: Q \to Y^*$, we usually call M a *generalized* sequential machine. We say that a sequential machine is *finite-*

Input string: Sequence of symbols from X	x_1	x_2	$\cdots$	x_n
State string: Sequence of functions f_u	f f_{x_1}	$f_{x_1 x_2}$	$\cdots$	$f_{x_1 x_2 \cdots x_n}$
Output string: Sequence of strings on Y^*	$f(\Lambda)$ $f(x_1)$	$f(x_1 x_2)$	$\cdots$	$f(x_1 x_2 \cdots x_n)$

$$F(x_1 x_2 \cdots x_n)$$

Figure 6.1   States of a sequential F.

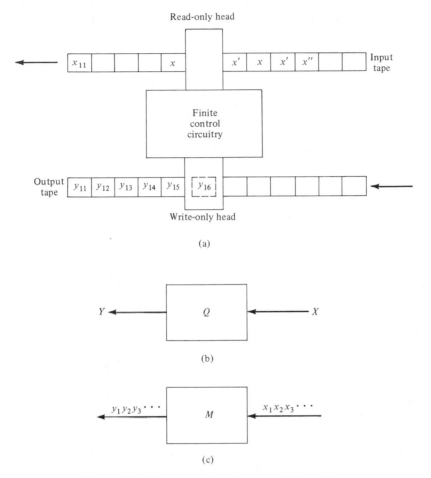

Figure 6.2   Three portraits of a sequential machine.

state if there are only finitely many distinct states. Thus a finite-state sequent-ial machine is just a finite automaton in the sense of Chapter 2.

Suppose that F has n states, and consider the effect of an input string of m inputs, for $m \geq n$. Then we have a sequence of $m + 1$ states

$$f_\Lambda, f_{x_1}, f_{x_1 x_2}, \ldots, f_{x_1 x_2 \cdots x_m}.$$

Since F has only n states, at least two of the above, say $f_{x_1 \cdots x_i}$ and $f_{x_1 \cdots x_j}$ with $i < j$, must be equal. Thus

3 Observation. *If F has only n states, and we apply a sequence of m > n inputs, at least one state will be repeated in the sequence of states entered by the machine.*

4 Observation. *It is clear that if F is finite-state with n states, then every F_u for u in X^*, is finite state with $\leq n$ states.*

5 Example. (i) Let $F: \{0,1\}^* \to \{0,1\}^*$ be the function with state $f: \{0,1\}^* \to \{0,1\}$ given by

$$f(x_1 x_2 \cdots x_n) = \begin{cases} 1 & \text{if an even number of } x_j s \text{ are 1,} \\ 0 & \text{if not.} \end{cases}$$

Then F is sequential, and is finite-state with only two states $f_0 = f$ and f_1 with the relations

$$f_1 = f_1 L_0 = f_0 L_1,$$
$$f_0 = f_1 L_1 = f_0 L_0.$$

that is, we only change state (parity) if the input is 1.

(ii) Let

$$G: \begin{Bmatrix} 0 & 0 & 1 & 1 \\ 0, & 1, & 0, & 1 \end{Bmatrix}^* \to \{0,1\}^* \text{ be the function.}$$

$$G\begin{bmatrix} u_1 & u_2 & u_3 & & u_n \\ v_1, & v_2, & v_3, & \cdots, & v_n \end{bmatrix} = \text{the } n \text{ lowest-order digits, with lowest-order digit first, of the binary expansion of the product of the binary numbers } u_n \cdots u_1 \text{ and } v_n \cdots v_1.$$

Then G is clearly sequential. However, G is *not* finite-state—we shall derive a contradiction from the assumption that G has a finite number n of internal states.

Suppose G had to multiply 2^n by itself fed to the machine as the string $\binom{0}{0}^n \binom{1}{1} \binom{0}{0}^n$ to yield 2^{2n}, coded as the string $0^{2n}1$. This would mean that after it had received its last nonzero input, G would have to print n more symbols, all zeros save for the last. But this would require that, setting $\hat{g} = fL_{(^0_0)^n}$ the state $\hat{g}_{(^0_0)^n}$ is different from each state $\hat{g}_{(^0_0)^j}$ for $0 \leq j < n$, contradicting **3** and **4**.

(iii) The string reversal function $H: \{0,1\}^* \to \{0,1\}^*$ with

$$H(x_1 x_2 \cdots x_n) = x_n \cdots x_2 x_1$$

is *not* sequential—for instance, $H(01)$ does not begin with the string $H(0)$.

Thus by restricting a machine to process a string one symbol at a time, or to preserve information about prior symbols by the present state from a *finite* set of possible states, we severely limit which functions from X^* to Y^* may be realized.

We are thus led to consider *off-line* machines, which may be simply defined as those functions that need not be sequential. We imagine that the whole string may be presented to the machine before any computation need take place. It is perhaps useful to think of the input string as read into a data structure that the machine may operate upon over a period of time, it usually being assumed that time is quantized, and that only a *finite* portion of the data structure is affected during each unit of time. (This condition will only be relaxed when we consider cellular automata in Chapter 7.)

There is a sense, then, in which we may view an on-line computation as treating an input string as distributed in time, whereas an off-line computation treats the string as distributed in space.

A major question, then, is this: Can we formally define a class of machines that can compute all partial functions $f: X^* \to Y^*$ which may be obtained by a well-defined machine when we place a finiteness condition not upon its memory but only upon its access to that memory? Since the notion of machine is informal in the last sentence, this amounts to finding a precise mathematical definition to replace our intuitive notion of an *effective procedure* for going (not always successfully, since the function may be partial) from a string of X^* to a string of Y^*.

The most important candidate for the notion of *effectively computable function* will be that of a *function computable by a Turing machine*. As researchers developed other theories of computation, they have found again and again that each computable function they specify can also be computed by Turing machine. This has led to the conviction that the notion of *Turing-computable* (and its equivalents) is indeed an adequate formalization of our intuitive notion of effectively computable. [See Chapter 1 of Kfoury, Arbib, and Moll (1982) for further discussion of this issue.] However, Turing machines often carry out their computation most inefficiently, and an important task of the automata theorist is to find more efficient automata to compute various classes of functions. We are now considering what effective computations are possible, without placing any bounds on the time or the storage space required to complete the computation. At this stage, we had better crystallize the idea of an *effective procedure*. There are certain computations for which there exist mechanical rules, or *algorithms*, e.g., the euclidean algorithm for finding the greatest common divisor of two integers. Certainly, any computation that can be carried out by a digital computer is governed by purely mechanical rules. We say, then, that there exists an effective procedure for carrying out these computations. There are many cases in which we do not really know how to write a program that would cause a given computer to carry out the desired computation, but we do have a strong intuitive feeling that a suitable effective procedure exists.

6.2 Turing Machines Introduced

Abstract automata theory may be said to start with the simultaneous publications of Turing (1936) and Post (1936), who gave independent—and equivalent—formulations of machines that could carry out any effective procedure provided they were adequately programmed. Of course, such a

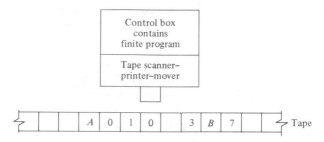

Figure 6.3   A Turing machine.

statement is informal—we cannot prove that the formally defined class of procedures implementable by Post or Turing machines will be the same as our intuitive hazy notions of effective procedure. But as we have already pointed out, this class has been proved equivalent to many other formal classes, and no one has produced a procedure that is intuitively effective but cannot be translated into a program for one of their machines.

The basic idea of the Post and Turing formalism is as follows (see Figure 6.3). The machine consist of

(i) a control box in which may be placed a *finite* program (i.e., which may be in one of a finite number of states);

(ii) a potentially infinite tape, divided lengthwise into squares (i.e., depending on our choice of mathematical fiction, we may consider the tape as comprising an infinite string of squares of which all but finitely many are blank, or as a finite tape to the ends of which arbitrarily many new squares may be added as required); and

(iii) a device for scanning, or printing on, one square of the tape at a time, and for moving along the tape, all under the command of the control box.

We start the machine with a finite *string* from X^* on the tape, and with a program in the control box. The symbol scanned and the instruction presently being executed (that is, the current state of the control box) uniquely determine what new symbol shall be printed on the square, how the head shall be moved, and what instruction is to be executed next (that is, what shall be the next state of the control box). Thus the control box of a Turing machine may be thought of as a finite-state sequential machine whose output can be either a halt instruction or a print-and-move instruction.

If and when the machine stops, the results of our computation, a new string from X^*, may be read off the tape.

We associate with a Turning machine Z a function $F_Z: X^* \to X^*$ by defining $F_Z(w)$ to be the expression printed on the tape when Z stops, if started in a specified initial state, say q_0, scanning the leftmost letter of the string w. If Z never stops after being so started, $F_Z(w)$ is to be left undefined.

Note that F_Z may be always defined, sometimes defined, or never defined.

Trivial examples of the three cases are, respectively, a machine that halts under all circumstances; a machine that halts if it scans a 1, but moves right if it scans any other symbol; and the machine that, no matter what state it is in and no matter what is sees, always moves right.

When the collection *Automata Studies* (Shannon and McCarthy, 1956) was published, automata theory emerged as a relatively autonomous discipline. Besides the "infinite-state" Turing–Post machines, much interest centered on finite-state sequential machines, which first arose not in the abstract form of our above discussion but, as we saw in Chapter 2, in connection with the input–output behavior of a McCulloch–Pitts net or an "isolated" Turing machine control box. The proof that any finite automaton can be implemented using neural nets shows that such nets can carry out the control operations of a Turing machine—providing, if you will, a formal "brain" for the formal machine that could carry out any effective procedure. These nets comprise synchronized elements, each capable of some boolean function, such as "and," "or," and "not." It was his knowledge of these networks that inspired von Neumann in establishing his logical design for digital computers with stored program, which is of basic importance to the present day. [In 1948, von Neumann (1951) added to the computational and logical questions of automata theory the new questions of construction and self-reproduction that we shall take up in Chapter 7.]

Turing's (1936) paper contains a charming "pseudopsychological" account of why we might expect any algorithms to be implementable by a suitable *A*-machine (his name for Turing machines). We reproduce excerpts from this below. Bear in mind that when Turing wrote this, "computer" meant a human who carried out computations.

> All arguments which can be given are bound to be, fundamentally, appeals to intuition [since the notion in intuitive] and for this reason, rather unsatisfactory mathematically . . . . Computing is normally done by writing certain symbols on paper. We may suppose this paper is divided into squares like a child's arithmetic book. In elementary arithmetic, the 2-dimensional character of the paper is sometimes used. But such a use is always avoidable, and I think that it will be agreed that the 2-dimensional character of paper is no essential of computation. I assume then that the computation is carried out on one-dimensional paper, i.e., on a tape divided into squares. I shall also suppose that the number of symbols which may be printed is finite. If we were to allow an infinity of symbols, then there would be symbols differing to an arbitrarily small extent . . . . It is always possible to use sequences of symbols in the place of single symbols . . . . The difference from our point of view between the single and compound symbols is that the compound symbols, if they are too lengthy, cannot be observed at one glance . . . . We cannot tell at one glance whether 9999999999 and 99999999999 are the same.
>
> The behavior of the computer at any moment is determined by the symbols which he is observing, and his "state of mind" at that moment. We may suppose that there is a bound B to the number of symbols on squares which the computer can observe at any moment. If he wishes to use more, he must use successive observations. We will also suppose that the number of

states of mind which need to be taken into account is finite. The reasons for
this are of the same character as those which restrict the number of symbols
. . . . Let us imagine that the operations performed by the computer are split
up into "simple operations," which are so elementary that it is not easy to
imagine them further divided. Every such operation consists of some change
of the physical system consisting of the computer and his tape. We know the
state of the system if we know the sequence of symbols on the tape, which of
those are observed by the computer (possibly with a special order), and the
state of mind of the computer. We may suppose that in a simple operation not
more than one symbol is altered, [and] . . . without loss of generality assume
that the squares whose symbols are changed are always "observed" squares.

Besides these changes of symbols, the simple operations must include
changes of distribution of observed squares. The new observed squares must
be immediately recognizable by the computer . . . . Let us say that each of the
new observed squares is within L squares of an immediately previously
observed square.

The simple operations must therefore include:

(a) Changes of the symbol on one of the oberved squares.
(b) Changes of one of the squares observed to another square within L
squares of one of the previously observed squares.

It may be that some of these changes necessarily involve a change of state
of mind . . . . The operation actually performed is determined . . . by the state
of mind of the computer and the observed symbols. In particular they
determine the state of mind of the computer after the operation is carried out.

We may now construct a machine to do the work of this computer. To
each state of mind of the computer corresponds an [internal state] of the
machine. The machine scans B squares corresponding to the B squares
observed by the computer. In any move the machine can change a symbol on
a scanned square, or can change any one of the scanned squares to another
square distant not more than L squares from one of the other scanned
squares.

The move which is done, and the succeeding [internal state] are determined
by the scanned symbol and the internal state. The machines just described do
not differ very essentially from [Turing machines] . . . [and so, a Turing
machine] can be constructed to compute . . . the sequence computed by the
computer.

If we think of Turing machines as actual physical devices, it is clear that the
input symbols could be configurations of holes on punched tape, patterns of
magnetization or handwritten characters, or the like, whereas the states could
be that of a clockwork, of a piece of electronic apparatus, or of an ingenious
hydraulic device. Such details are irrelevant to our study, and so if there are m
inputs in X, we shall feel free to refer to them as $x_0, x_1, \ldots, x_{m-1}$ without
feeling impelled to provide further specification; and if there are n states, we
shall similarly find it useful to label them $q_0, q_1, \ldots, q_{n-1}$. For each abstract
description there are many implementations—depending, e.g., on whether we
interpret x_3 as a 0 or a 1, as a pattern of neural activity, or a configuration of
holes in a punched tape—but it should be clear that, from the information-

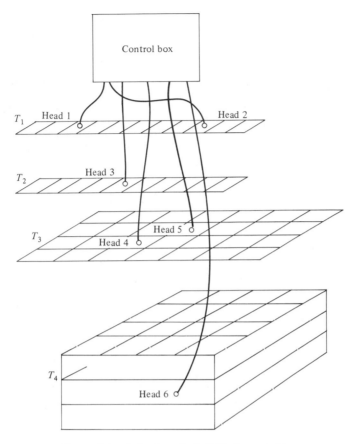

Figure 6.4 A "polycephalic" Turing machine.

processing point of view, there is a very real sense in which all these machines may be considered the *same* machine.

Computation more efficient than that of ordinary Turing machines can be obtained simply by allowing the Turing machine to have several tapes—and these not necessarily one-dimensional—each acted on by one or more heads that report back to a single control box which coordinates their printing and moving on their respective tapes (Figure 6.4). Any job that can be done by such a "polycephalic" Turing machine can be simulated by an ordinary Turing machine—thus reassuring ourselves of the breadth of the notion of a Turing-machine-computable (henceforth: TM-computable) function [which, incidentally, is coextensive with the notion of partial recursive function introduced by the logicians in the 1930s as an alternative formalization of the effectively computable functions (Kfoury, Moll, and Arbib, 1982, Chapter 9)]. Furthermore, these polycephalic machines do indeed give us the expected gain in efficiency. They not only have the great virtue of being more efficient in that they take less time to complete a computation, but also are easier to write programs for. With these "polycephalic" machines we have a realistic model of computers—a virtue we do not claim for the ordinary Turing machine, useful though it is in allowing us to construct a theory of the computable. With these machines we are also made aware that there is no reason to restrict automata theory to functions of the form $F: X^* \rightarrow Y^*$. With multidimensional tapes, our automata may as well process planar or higher-dimensional configurations as the linear strings of X^*, as in our study of pattern recognition in Chapter 4, and in our study of cellular automata in Chapter 7.

The reader should note that any present-day computer is really a generalized Turing machine: the input system corresponds to a read-only tape, the output system corresponds to a write-only tape, the tape units (although of only finite capacity) correspond to one-dimensional, one-head TM tapes, and addressable words of the memory correspond to TM tapes only one square long. (The machine can read and write but cannot move on these tapes.)

In fact it can be proved that the functions computable by these machines are all computable by the conventional one-head, one-tape, one-dimensional machine [see Arbib (1969, Section 4.2) for the formal definitions, and a description of the simulation process]. What we do gain by using more complicated machines is speed and efficiency. For instance, Barzdin (1965) has proved that no machine with one head can tell whether or not a word is symmetric, that is, is a palindrome, in less than a time that increases proportionately with the *square* of the length of the word; whereas it is easy to see if it has two heads, we may move the heads to the opposite end of the tape and have them move in to the middle, reporting back to the control, so that when they meet, in a time that only goes up with the length of the tape, thay will have been able to tell whether or not the string is a palindrome.

Here is an extremely important question of automata theory: Given a problem, what is the computing structure best suited to it? For instance, I presume that one could rigorously prove that adding extra heads to the two-head machine could not make its recognition of palindromes any more efficient, and so the two-head structure is one most suited for a device for recognizing palindromes. We may note too that authors such as Wang (1957) and Minsky (1967) have shown that even simpler instruction sets than that of the Turing machine suffice to give us all effective procedures; the price one pays is even less efficiency than that of the Turing machine. We thus see the need to understand the tradeoff between complexity of the "genetic" structures and the complexity of the computations that take place within those structures.

Let us indicate the power—and the inefficiency—of ordinary Turing machines by outlining how we might program one to handle the example of string reversal.

1. Start in initial state at the left-hand end of the tape, which bears a special marker followed by the string to be reversed:

$$\overset{\downarrow}{*} x_1 x_2 \cdots x_{n-1} x_n.$$

2. Move right to the first blank. Go back one square. Note the symbol there, erase it, move one square right and print it:

$$* x_1 x_2 \cdots x_{n-1} b \overset{\downarrow}{x_n}$$

(where we use b to remind us of the presence of a blank square).
3. Move left until you first scan a nonblank symbol immediately after scanning a blank. If the symbol is $*$ go to 4. If the symbol is not $*$,

$$* x_1 x_2 \cdots \overset{\downarrow}{x_j} b \cdots b x_n x_{n-1} \cdots x_{j+1},$$

note the symbol there, erase it, and move right until you first scan a blank immediately after scanning a nonblank symbol, and print the noted symbol

$$* x_1 x_2 \cdots x_{j-1} b b \cdots b x_n x_{n-1} \cdots x_{j+1} \overset{\downarrow}{x_j}$$

and repeat all the instructions in 3.
4. If, in moving left, the first nonblank symbol to the left of a blank is $*$,

$$\overset{\downarrow}{*} b b \cdots b b x_n x_{n-1} \cdots x_2 x_1$$

erase $*$ and halt. The nonblank portion of the tape contains the reversed string.

An interesting point about this computation is that we only need about twice as much tape as that on which the data are printed originally. And so another question will be: If we do have a computation that is off-line, requiring us to present the total data in a structure on which the machine can operate, how much auxiliary storage will be required to complete the computation? Again there will often be a tradeoff between time and storage requirements, some of which are discussed by Arbib (1969, Chapter7).

Since each Turing machine is described by a finite list of instructions, it is easy to show that we may *effectively* enumerate the Turing machines

$$Z_0, Z_1, Z_2, Z_3, \ldots$$

so that, given n we may effectively find Z_n and given the list of instructions for Z we may effectively find the n for which $Z = Z_n$. For example, we may think of the description of each machine as a long word (whose characters may include blanks and punctuation marks). We can then list these programs/words in order of increasing length, while—having fixed an "alphabetical order" for all the characters—we can list the programs/words of a given length in lexicographic order.

This implies that we can effectively enumerate all computable (more precisely, Turing machine computable) functions as:

$$\phi_0, \phi_1, \phi_2, \phi_3, \ldots$$

simply by setting $\phi_n = F_{Z_n}$. Such effective enumeration lies at the heart of much of our study of the computable. For example, if we say that ϕ_n is *total* if $\phi_n(w)$ is defined for all w, we might ask:

Does there exist an effective enumeration of all and only the total computable function, i.e., a total computable function h such that ϕ_n is total if and only if $n = h(m)$ for some m (identifying a string with a suitable number that encodes it)?

The answer is "NO," for if such an h existed, we could define f by

$$f(n) = \phi_{h(n)}(n) + 1.$$

Then f would be total computable, and so $f = \phi_{h(m)}$ for some m. Then $\phi_{h(m)}(m) = \phi_{h(m)}(m) + 1$ a contradiction!

Suppose we could effectively tell, given n, whether or not the function ϕ_n was total, i.e., *suppose* there existed a computable function g such that

$$g(n) = \begin{cases} 1 & \text{if } \phi_n \text{ is total,} \\ 0 & \text{if not.} \end{cases}$$

Pick any number n_0 such that ϕ_{n_0} is total, and define the total function h by

$$h(n) = \begin{cases} n & \text{if } g(n) = 1, \\ n_0 & \text{if } g(n) = 0. \end{cases}$$

If g is computable, *then* h would be computable. But h has the property that ϕ_n is total if and only if $n = h(m)$ for some m. However, we have seen that such an h cannot be computable, and so we conclude that g *cannot* be computable either.

This is just one example of the many things we can prove to be undecidable by any effective procedure. To say that we cannot effectively tell that f_n is total is just the same as saying that we cannot tell effectively whether Z_n will stop computing no matter what tape it is started on. We may thus say that *the halting problem for Turing machines is unsolvable.*

One of the most interesting results in Turing's original paper 1936 was that there exists a *universal* Turing machine.

1 Theorem. *There exists a Turing machine U such that given the encoding $e(Z)$ of the program of a Turing machine Z and a data string w on its tape, it will proceed to simulate the computation of Z on w, halting if and only if $F_Z(w)$ is defined, in which case it will halt with $e(Z)$, $F_Z(w)$ on its tape:*

$$F_U(e(Z),w) = e(Z),F_Z(w).$$

PROOF OUTLINE. The idea of the proof is that U, in simulating Z, is to have its tape divided into two halves: the left-hand half contains the list of coded instructions with a pointer indicating which state Z would be in at this stage of the simulation; while the right-hand half contains the data string that Z would have on its tape, with a pointer indicating which square of its tape Z would be scanning. The strategy is for the control box of U to read an instruction, use the next-state portion of it to reposition the current-state marker (this involves some subtleties, since Z may have far more states than U), and then move right until it finds the current-symbol marker to carry out the designated tape manipulation (problems arise here if new data are to be printed on a new square to the left of Z's tape) before moving left to find the current-state marker and initiate the cycle of simulation once more. ☐

The reader may wish to try filling in the details before turning to the elegant construction of Minsky (1967).

With this result before us, we can prove one of the most surprising theorems about effective computations: that any enumeration of computable functions is redundant in any way that we can effectively specify.

2 The Recursion Theorem. *Let* $\phi_0, \phi_1, \phi_2, \ldots$ *be any effective enumeration of the computable functions, and let h be any total TM-computable function. Then there exists a number i_0 such that*

$$\phi_{i_0} = \phi_{h(i_0)}.$$

We call i_0 a fixed point *for h.*

PROOF. Consider the following functions:

$\alpha(z, x) = z, z, x$ This just involves copying the first argument and so is effectively computable.

$\beta(z, y, x) = \phi_z(y), x$ This involves applying the universal Turing machine to the first two arguments, while preserving the third, and so is effectively computable.

$\gamma(w, x) = \phi_w(x)$ This is just the universal Turing machine function.

Thus

$$\gamma \circ \beta \circ \alpha(z, x) = \phi_{\phi_z(z)}(x)$$

is an effectively computable function of z and x, and so given z we may effectively find an index $g(z)$ for the function $\gamma \circ \beta \circ \alpha(z, .)$. We thus have for all z and x

$$\phi_{g(z)}(x) = \phi_{\phi_z(z)}(x).$$

Let then v be the index of the total function $h \circ g$:

$$\phi_v(x) = h(g(x)).$$

Thus

$$\phi_{\phi_v(v)}(x) = \phi_{h(g(v))}(x).$$

But by definition of g

$$\phi_{\phi_v(v)}(x) = \phi_{g(v)}(x).$$

Thus, $g(v)$ is the desired index i_0 such that

$$\phi_{i_0} = \phi_{h(i_0)}.$$ □

An interesting application of this result is the following, due to Lee (1963).

3 Theorem. *There exists a self-describing Turing machine; that is, there exists an integer n_0 such that Z_{n_0} will print out its own "description" n_0 on an initially blank tape:*

$$\phi_{n_0}(\Lambda) = n_0.$$

PROOF. Define h to be the function such that for each n, $h(n)$ equals an index of a machine that, when started with blank tape, prints out n and halts, i.e., such that $\phi_{h(n)}(\Lambda) = n$. Then let n_0 be a fixed point for h:

$$\phi_{n_0}(\Lambda) = \phi_{h(n_0)}(\Lambda) = n_0.$$ □

6.3 Recursive and Recursively Enumerable Sets

Now that we have a fairly good understanding of the partial functions F: $X^* \to X^*$ which can be computed by Turing machines, we wish to study the subsets of X^* characterized by these functions. In what follows, when we are discussing a fixed alphabet X, we shall sometimes use an integer n to refer to a string of X^*, it being understood that if $|X| = k$, then by n we really mean the string on X which seves as k-akic coding of n (under some fixed correspondence between $\{1, 2, \ldots, k\}$ and X).

1 Definition.

(i) (Turing machines as *acceptors*.) A subset R of X^* is *recursive* just in case its characteristic funtion

$$\chi_R(n) = \begin{cases} 1 & \text{if } n \in R, \\ 0 & \text{if } n \notin R, \end{cases}$$

is computable.

(ii) (Turing machines as *generators*.) A subset R of X^* is *recursively enumerable* just in case it is the empty set $\varnothing^{\dagger}$ or is the range of some total computable function f

$$R = \{f(n) | n \in X^*\}.$$

[†] Note that $\varnothing$ is the range of a partial computable function, namely, that undefined for all inputs. We shall prove that a set is in fact recursively enumerable iff it is the range of a partial computable function.

We now talk about sets of numbers—the reader should make a mental translation to the general case. Since a set R is an actual collection of numbers we can order it (enumerate it) in many different ways. To say a set is recursively enumerable is thus to say that for *at least one enumeration*, the successive values of a computable function f correspond to the successive (perhaps with repetitions) elements of R in the enumeration. Pick such an enumeration by some computable function f. Now ask yourself, of some value n: Does or does not n belong to R? If we generate the successive values of f: $f(0)$, (1), $f(2)$, ..., we will eventually reach any member of R, and so effectively tell if n *does* belong to R. However, if n does *not* belong to R, we can never tell by this method, for even if we generate a billion elements of R, we cannot be sure that n is not the billion and first. So given a set R and a function f which recursively enumerates it, $R = \{f(0), f(1), f(2), \ldots\}$, one *cannot* use f directly to see if R is *recursive*, but one might be able to use it indirectly, via knowledge of a program that generates it, to prove that there is a total *computable* function, $g: N \to \{0, 1\}$, such that

$$g(n) = 1 \quad \text{if and only if } n = f(m) \text{ for some } m.$$

If we can prove that no such recursive g exists, then we deduce that R is *not* recursive. The distinction I am trying to emphasize, then, is that the Turing machine which enumerates a set is quite distinct from the Turing machine (if there is one) which checks membership in the set.

The set of squares of integers is *recursively enumerable*—we take 1, 2, 3, ... in turn and square them. It is also *recursive*   given any integer, we decompose it into its prime factors and then tell easily whether or not it is square. We shall see that every recursive set is recursively enumerable, but that the converse is not true. The latter result is deep.

No one denies that the design of a Turing machine may require enormous intellectual effort—the point is that a problem is called "recursively solvable" if this intellectual effort results in a finite set of rules that can thereafter be applied *mechanically* to solve the problem, no matter what form the parameters may take. The next point is that the procedure is recursive if we can guarantee that it will terminate whenever an answer is defined—but *recursiveness* is no guarantee of practicability, for to say that a process will terminate is no guarantee that it will terminate in a reasonable span of time. (It is this difficulty that leads one to ask of a function not merely "is it recursive?" but also "how difficult is it?")

We reiterate the hypothesis (a variant of which if often called *Church's thesis*), following Turing in his original 1936 paper: *The informal intuitive notion of an effective procedure on sequences of symbols is identical with our precise concept of one that may be executed by a Turing machine.* Let us then recouch our definitions in our intuitive language of effective procedures:

2. A *function* is called *computable* if there exists an effective procedure for computing it (computation).

A *set* is *recursive* if there exists an effective procedure for telling whether or not an element belongs to it (decision).

A *set* is *recursively enumerable* if there exists an effective procedure for generating its elements, one after another (generation).

We now have two sets of definitions, formal and informal. If a function is recursive in the formal sense, it is certainly recursive in the informal sense— our effective procedure is simply to compute the function with the Turing machine given to us by the formal definition. Similarly, a set that is recursive (or recursively enumerable) in the formal sense must be recursive (or recursively enumerable) in the informal sense. It is the converse statement (e.g., that if a set is recursive in the informal sense, then it is recursive in the formal sense) that constitutes, for us, Turing's hypothesis. We now prove theorems informally on recursive and recursively enumerable sets. Of course, when a formal proof does not involve many finicky details, we shall not hesitate to give it in full.

3 Theorem. *If R and S are recursively enumerable sets, then so are $R \cap S$ and $R \cup S$.*

PROOF. The cases in which R or S is null are trivially true. Otherwise, let $R = \{f(n)|n \in N\}$, $S = \{g(n)|n \in N\}$ for total computable f and g. Let

$$h(n) = \begin{cases} f(m) & \text{if } n = 2m + 1, \\ g(m) & \text{if } n = 2m. \end{cases}$$

Then h is clearly computable and $R \cup S = \{h(n)|n \in N\}$ and so is recursively enumerable.

If $R \cap S = \varnothing$, it is certainly recursively enumerable. If $R \cap S \neq \varnothing$, there exists $m_0 \in R \cap S$ (although we may not have an effective means of finding it— see the discussion following Theorem 7). Let us enumerate the pairs of integers $(0, 0)$, $(1, 0)$, $(0, 1)$, $(2, 0)$, $(1, 1)$, $(0, 2) \ldots$ (see Figure 6.5 for the pattern), and define total computable functions p_1, $p_2 : N \to N$ by letting $(p_1(n), p_2(n))$

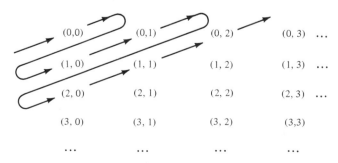

Figure 6.5   An effective enumeration of all pairs of natural numbers.

denote the nth pair in the series. Now define the total computable function k by

$$k(n) = \begin{cases} p_1(n) & \text{if } f(p_1(n)) = g(p_2(n)), \\ m_0 & \text{otherwise.} \end{cases}$$

Then clearly $R \cap S = \{k(n) | n \in N\}$. □

4 Observation. $S \subset X^*$ *is recursive iff its complement* $\bar{S} = X^* - S$ *is recursive.*

5 Theorem. *A set S is recursive iff both S and $\bar{S}$ are recursively enumerable.*

PROOF. Let S be recursive, with g a total computable function such that

$$g(n) = \begin{cases} 1 & \text{if } n \in S, \\ 0 & \text{if } n \notin S. \end{cases}$$

Let n_0 be the least element of S. (We may assume such exists, since the theorem clearly holds for the empty set.) Then define the effectively computable function f as follows:

$$f(0) = n_0,$$

$$f(n) = \begin{cases} f(n-1) & \text{if } g(n) = 0, \\ n & \text{if } g(n) = 1. \end{cases}$$

Then $S = \{f(n) | n \in N\}$ and so is recursively enumerable—and similarly for $\bar{S}$. Conversely, suppose S and $\bar{S}$ are recursively enumerable. Then

$$S = \{f(n) | n \in n\} \quad \text{and} \quad \bar{S} = \{h(n) | n \in N\}$$

for some total computable functions f and h, and given any $m \in N$, we know m is either an $f(n)$ for some n, or an $h(n)$ for some n, but not both. So given m, simply generate the sequence $f(1), h(1), f(2), h(2), f(3), \ldots$ until we encounter m. If it is an $f(n)$, set $g(m) = 1$; if it is an $h(n)$, set $g(m) = 0$. Then

$$g(m) = \begin{cases} 0 & \text{if } m \notin S \\ 1 & \text{if } m \in S \end{cases}$$

and so S is recursive. □

6 Observations.

(i) *A subset of N is recursive iff it can be recursively enumerated in order of magnitude.*
(ii) *Any finite set is recursive.*
(iii) *If S is finite and R is recursive (recursively enumerable), then $R \cup S$ is recursive (recursively enumerable).*

We may relax the condition that the function used to enumerate a recursively enumerable set be total:

7 Theorem. *A set is recursively enumerable iff it is the range of a (partial) computable function.*

PROOF. We have to show that $\{\phi(n)|n \in N\}$ is recursively enumerable, even if the computable function ϕ is *not* total. If ϕ is completely undefined, then its range is $\varnothing$, and we are done. If not, pick some n_0 from its range.

Let Z be a Turing machine that computes ϕ. Let

$$F(n, m) = \begin{cases} 1 & \text{if } Z \text{ takes } m \text{ steps to compute } \phi(n), \\ 0 & \text{if not.} \end{cases}$$

Now, F is total recursive—to compute it, we simply start Z with n on its tape, and run it for m steps. If it then stops, $F(n, m)$ is 1; otherwise, 0. Let us now define the function g by the equalities

$$g(n) = \begin{cases} \phi(x) & \text{if } p_1(n) = x, p_2(n) = y, \text{ and } F(x, y) = 1, \\ n_0 & \text{otherwise.} \end{cases}$$

Then g *is* total computable, and $\{g(n)|n \in N\} = \{\phi(n)|n \in N\}$, which is thus recursively enumerable.   □

The reader should ponder the fact that *there many be an effective procedure for solving a problem—but no effective procedure for finding a solution procedure.* This is well illustrated by the above proof. If we generate a set by a partial recursive function, we run the risk that the output may be undefined— so we adopt the "run for y steps with input x" strategy to ensure that computation will halt for every input n—which we decode by solving the equation $(x, y) = (p_1(n), p_2(n))$. This transition is effective, save for one catch—we must generate an element of the set even if the original computation on input x would not have taken y steps. To plug this gap, we fix on a single element n_0 of the range of the original recursive function. This we can obtain, if we are patient enough, by our "run for y steps with input x" strategy till we first obtain an output—*so long as the range of our partial function is nonempty.* But to tell whether or not it is empty is a problem akin to the halting problem and thus effectively unsolvable—as the reader may readily see. Hence, if we *know* our set is nonempty, we can find an algorithm of the desired guaranteed-to-terminate variety.

8 Theorem. *The recursively enumerable sets may be effectively enumerated:* S_0, $S_1, \ldots S_n, \ldots$

PROOF. Let $Z_0, Z_1, \ldots$ be an effective enumeration of Turing machines. Let ϕ_n be the partial recursive function of one argument computed by Z_n. Let $S_n = \{\phi_n(m)|m \in N\}$.   □

9 Theorem. *There exists a recursively enumerable set that is not recursive.*

PROOF. By Theorem 5 we must exhibit a recursively enumerable set K for which $\bar{K}$ is not recursively enumerable.

Let us define the set K to be $\{n \mid n \in S_n\}$. Thus

$$n \in K \Leftrightarrow n \in S_n,$$

where $\{S_0, S_1, \dots\}$ is an effective enumeration of the recursively enumerable sets.

K is recursively enumerable, for, if we define ϕ by

$$\phi(n) = \begin{cases} y & \text{if } (x, y) = (p_1(n), p_2(n)) \text{ and } y = \phi_y(x), \\ \text{undefined} & \text{if not,} \end{cases}$$

then $K = \{\phi(n) \mid n \in N\}$.

Finally, we show that $\bar{K}$ is not recursively enumerable. For were $\bar{K}$ recursively enumerable, we would have $\bar{K} = S_{n_0}$ for some n_0. But then

$$n_0 \in K \Leftrightarrow n_0 \in S_{n_0} \Leftrightarrow n_0 \in \bar{K}$$

a contradiction.   □

It is instructive to contrast the above with the following "proof": "Since every recursive set is recursively enumerable, all recursive sets certainly ocur in $S_1, S_2, \dots$. If it were true that all sets $S_1, S_2, \dots$ were recursive, then the set $\bar{K} = \{n \mid n \notin S_n\}$, which is not in the enumeration, would be recursive—for, S_n being recursive, we can effectively tell whether or not $n \in \bar{K}$ by testing whether or not $n \in S_n$. Contradiction—and so $S_1, S_2, \dots$ contains nonrecursive sets—and hence there are recursively enumerable sets which are not recursive."

This proof is invalid. The reader is invited to figure out why before reading the rest of this paragraph. Why? Because, by reasoning similar to that following Theorem 7 we cannot in general go effectively from the Turing machine Z_n which generates S_n to a decision procedure that tells whether or not a number belongs to S_n—even if we know the set is recursive. But for the above proof to work, we must show that $\bar{K}$ is recursive by *effectively* obtaining for each n a decision procedure to test its membership in S_n. What the above argument *does* prove, however, is that any effective enumeration of decision procedures for recursive sets cannot include decision procedures for all recursive sets.

References for Chapter 6

Arbib, M.A., 1969, *Theories of Abstract Automata*, Prentice-Hall.

Barzdin, Y.M., 1965, Complexity of recognition of symmetry in Turing machines, *Problemy Kibernetiki* **15**.

Kfoury, A.J., Moll, R.N., and Arbib, M.A., 1982, *A Programming Approach to Computability*, Springer-Verlag.

Lee, C.H., 1963, A Turing machine which prints its own code script, in *Proc. Symp. Math. Theory of Automata*. Polytechnic Press, pp. 155–64 (Vol. XII of the Microwave Research Institute Symposia Series).

McCulloch, W., and Pitts, W., 1943, A logical calculus of the ideas immanent in nervous activity, *Bull. Math. Biophys.* **5**, 115–133.

Minsky, M., 1967, *Computation: Finite and Infinite Machines*, Prentice-Hall.

Moll, R.N., Arbib, M.A., and Kfoury, A.J., 1987, *An Introduction to Formal Language Theory*, Springer-Verlag.

Post, E.L., 1936, Finite combinatory processes—Formulation I, *Symbolic Logic* **I**, 103–105.

Shannon, C.E., and McCarthy, J., Eds., 1956, *Automata Studies*, Princeton University Press.

Turing, A.M., 1936, On computable numbers, with an application to the Entscheidungsproblem, *Proc. London Math. Soc.*, Ser. 2 **42**, 230–265, with a correction, *ibid.*, **43** (1936–7), 544–546.

von Neumann, J., 1951, The general and logical theory of automata, in *Cerebral Mechanisms of Behavior: The Hixon Symposium* (L.A. Jeffress, Ed.), Wiley, pp. 1–32.

Wang H., 1957, A variant to Turing's theory of computing machines, *J.A.C.M.* **4**: 63–92.

CHAPTER 7

Automata that Construct as well as Compute

In Section 7.1, we introduce the notion of a cellular (or tessellation) automaton, and show how to embed CT-machines within the cellular array. These have the computational power of Turing machines (thus the T) as well as the ability to Construct other CT-machines (thus the C). We then prove results about universal constructors and self-reproducing machines. In Section 7.2, we further formalize the notion of cellular automata, and prove that there exist Garden-of-Eden configurations that cannot arise in the cellular array without intervention of an external agency. Finally, Section 7.3 discusses differences between the type of self-reproduction formalized here and the growth processes of embryology.

7.1 Self-Reproducing Automata

Von Neumann (1951) noted that when machines built other machines, there was a degradation in complexity (an assembly line is more complicated than what it produces), whereas the offspring of an animal seems generally to be at least as complex as the parent, with complexity increasing in the long range of evolution. Von Neumann asked whether there was an immutable difference here, or whether one could design "self-reproducing machines" that could produce other machines without a degradation in complexity. Such a question suggests that we study automata theory not from that aspect emphasized in earlier chapters, namely, processing of input patterns to yield outputs, but rather a more constructional approach in which we study how information can be read out and processed to control the growth and change in structure of an automaton.

When we have an adaptive network that is changing certain parameters with time, it may *not* be useful to make the distinction between growth processes and parameter adjustment processes (see Chapters 4 and 5). If we design a network to carry out a computation, there are at least two ways to build it. One is to connect each component to all other components, and then adjust the coupling coefficients through computation. One then might claim that information processing did not change structure, even though in some cases is would lead to a zero coupling coefficient, which would be equivalent to no connection at all (see Chapter 3). Alternatively, one might use a process that starts with very few connections. Then if we compute that one component should, in fact, influence another that it has not before, we may use this to "grow" a connection between the two components. Thus, as we come to understand in more detail how genes help to control embryological development, we may hope to be able to understand, by extrapolation of those processes, something of the mechanisms that underlie learning and memory.

To return to von Neumann's question, let us see how we might describe a complicated machine that can produce something as complicated as itself. In Section 7.3 we shall then analyze some of the many ways in which the reproduction of such an automaton differs from biological reproduction. However, for now we shall study "pure" automata theory with little explicit concern for it applicability.

Let us first observe that the question: "Is there a machine that, if set loose in a component-rich environment, will form components into a copy of itself?" can be given a trivial answer, as in the domino example of Figure 7.1. The problem is that the "reproducing machine," the falling domino, is too simple to be of interest. To avoid this triviality, we consider the following artificial, but at least automata-theoretic, question: "Can we design an automaton able to simulate that Turing machine, which also can reproduce itself?"

To formalize this (admittedly nonembryological) problem, consider an infinite "chess board," with each square either empty or containing a single component. Each component can be in one of various states, and we think of an organism as represented by a group of cells, collected together somewhere in the plane (see Figure 7.2). We are thus talking of regions in what we call a *tessellation automaton* or *cellular automaton*.

A mathematical simplification: We have said that any square of the board may be empty or contain some component, say of type $C_j (1 \leq j \leq N)$, in some state, say q_i. We may lump these $N + 1$ alternatives into one supercom-

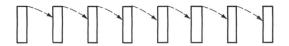

Figure 7.1   Trivial self-reproduction. A domino on edge is the basic component. We stand dominoes in a chain, as shown, and let the automaton we want to reproduce be a *falling* domino. A falling domino knocks down its neighbor, and thus "reproduces." *Falling* is propagated down the chain.

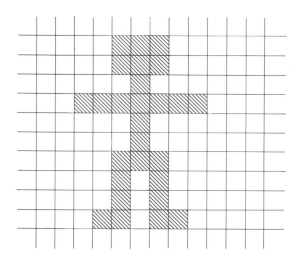

Figure 7.2   An "organism" of active (hatched) cells embedded in a chess-board array.

ponent $\hat{C}$, which has one more state than the total number of states of the N components. For mathematical purposes, it is easier to think of there being a copy of one fixed component in every cell, so that rather than study the kinematics of components moving around in the plane we look at a more tractable process of how an array of identical components $\hat{C}$, consisting initially of an activated subarray with the remaining cells in the passive state, passes information to compute and to "construct" new configurations.

Von Neumann was able to show that with a 29-state cell "supercomponent," he could set up a simulation of a complex Turing-type machine that, besides being able to carry out computations on its tape, would also be able to "reproduce" itself. The 29 states could be seen as several states corresponding to an OR gate, several states corresponding to an AND gate, several states corresponding to different types of transmission line, and so forth. Von Neumann's proof was not completed at the time of his death, but the manuscript he left was edited by Arthur Burks and published as *The Theory of Self-Reproducing Automata* (von Neumann, 1966). The proof is over 100 pages long. The price we pay for simple components is a complex program. To take an analogy from computer programming, it is like trying to program in machine language, rather than in an appropriate assembly language. In biological terms, we might say that it is like trying to understand a complicated organism directly in terms of macromolecules, rather than via the intermediary of cellular structure.

Turing's result that there exists a universal computing machine suggested to von Neumann that there might be a universal construction machine A, which, when furnished with a suitable description I_N of any appropriate automaton N, would construct a copy of N (see Figure 7.3). In what follows, all automata for whose construction we use A will share with A the property

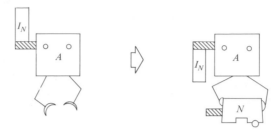

Figure 7.3   Fanciful description of universal constructor A.

of having a place where an instruction I can be inserted. We may thus talk of "inserting given instruction I into a given automaton."

If the automaton A has description I_A inserted into it, it will proceed to construct a copy of A. However, A is *not* self-reproducing, for A with appended description I_A produces A with I_A; it is as if a cell had split in two with only one of the daughter cells containing the genetic message. Adding a description of I_A to I_A does not help; now $A + I_{A+I_A}$ produces $A + I_A$, and we seem to be in danger of an infinite regress. Such a consideration suggested to von Neumann that the correct strategy might involve "duplication of the genetic material." He thus introduced an automaton B that can make a copy of any instruction I with which it is furnished, I being an aggregate of elementary parts, and B just being a "copier" (see Figure 7.4). Next, a third automaton C will insert the copy of I into the automaton constructed by A. Finally, C will separate this construction from the system $A + B + C$ and "turn it loose" as an independent entity.

Let us then denote the total aggregate $A + B + C$ by D. In order to function, the aggregate D must have an instruction I inserted into A. Let I_D be the description of D, and let E be D with I_D inserted into A. Then E is self-reproducing and no vicious circle is involved, since D exists before we have to define the instructions of I_D.

We thus see that once we can prove the existence of a universal constructor for automata constructed of a given set of components, the logic required to proceed to a self-reproducing automaton is very simple, although there is something somewhat whimsical in the idea of a universal constructor, as if a mother could have offspring of any species, depending only on the father. While this may be appropriate to Greek myths, it does seem inappropriate to biological modeling—unless we think of experiments in which cytoplasm plays host to a transplanted nucleus. It would seem that animals of the same species are such that (the cytoplasm is compatible for readout of the code of any member and) the "construction program" is structured in the same way for all members at the level which treats the subroutines as single constructions. Members of the species then seem to differ in the details of these subroutines. With the added mechanism of dominance, we see why sexual recombination always yields valid programs within a species, but may yield meaningless programs between species when two different program structures

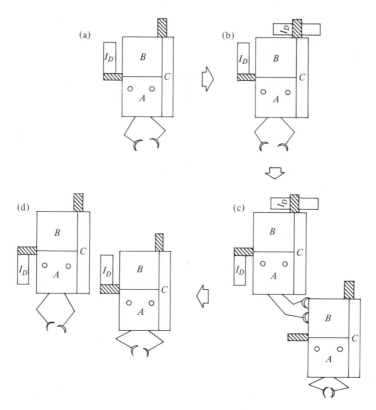

Figure 7.4   Fanciful description of a self-reproducing machine.

are paired in what is computationally a random fashion. This also suggests why the notion of a universal constructor seems so whimsical *biologically*, although perhaps cytoplasm may be a universal constructor for the combined genetic message of mother and father; however, I suspect that cytoplasm itself has more evolutionary memory than we normally ascribe to it.

Our concern now is to examine the difficulties involved in actually providing a universal constructor. A Turing machine is only required to carry out elementary logical manipulations on its tape, sensing symbols, moving the tape, printing symbols, and carrying out elementary logical operations. A universal *computer* only has to carry out the same operations, but a universal *constructor* must also be able to recognize components, move them around, manipulate them, join them together. Thus, presumably, constructors of Turing machines require more components than do Turing machines themselves. We are immediately confronted with the possibility of another infinite regress. Given a set of components C_1, to construct machines that build all the automata made from components of C_1, we may need a bigger set of components C_2. To build all machines constructed of components from C_2, we may need machines put together from a bigger set of components C_3. The question is: "Is there a *fixed point*? Can we find a set of components C such that all automata built from components of C can be constructed by automata built from the same set C?" This *fixed-point problem for components* is the fundamental problem in the theory of self-reproducing automata. Once we have found a set of components C in which for each automaton A there can be found an automaton $c(A)$ that constructs A, it turns out to be a fairly routine matter to prove the existence of a universal constructor. We then know from von Neumann that it is a simple matter to prove *the construction fixed-point theorem*, namely, that there exists a self-reproducing machine U which can construct a copy of U. There have been several procedures following von Neumann's to exhibit a set of components that satisfy the component fixed-point theorem. Thatcher (1970) used the same 29-state components as von Neumann, but gave a more elegant construction of perhaps half the length. Codd (1965), with remarkable ingenuity and interaction with a computer, showed that a construction similar to von Neumann's could go through using components with only eight states. I showed (Arbib, 1966, 1967) that the construction could be done with great simplicity, in a matter of eight pages, if one allowed the use of much more complicated components. My rationalization for this use of complex components was that if one wishes to understand complex organisms, one should adopt a hierarchical approach, seeing how the organism is built up from cells, rather than from macromolecules.

Rather than go into any details of my construction I shall just present a few pictures that give some idea of the basic notions involved. We are to imagine a *CT* machine (Construction and Turing machine) that under the control of a program in its logic box can read and write on a one-dimensional tape in just the way a Turing machine does, and that can write but not read on a two-dimensional tape (Figure 7.5) which serves as a *construction area* where the

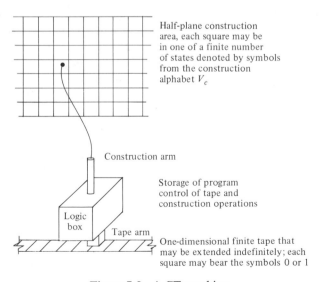

Figure 7.5   A CT-machine.

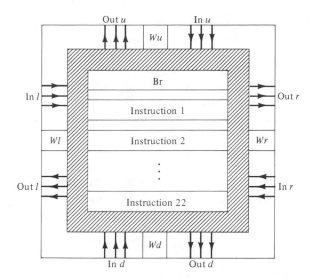

Figure 7.6   The basic module.

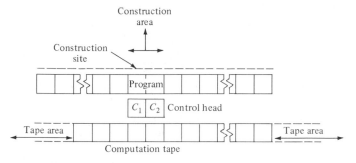

Figure 7.7   Overall plan of embedded CT machine.

writing of a symbol is equivalent to the placing of a component. Our task is to find a set of components from which we can build tape, logic box, and construction area. Such a component as shown in Figure 7.6 is a finite-state module that can contain up to 22 instructions from a rather limited instruction set. These cells are embedded in a two-dimensional plane in the manner of Figure 7.2 An automaton is then represented by an activated configuration of these cells. The little boxes marked W in Figure 7.6 are *weld registers*, which enable a number of squares to be "welded" into a one-dimensional tape in such a way that when any one square of that tape is instructed to move, all cells will move in the indicated direction. The assumption of such a weld operation greatly simplifies our programming.

Figure 7.7 shows the overall plan of an embedded CT machine. The logic box has been broken into two pieces, a one-dimensional tape that contains the program and two cells that form a control head. The idea is that on activation by the control head, squares of the program tape may either be used to guide the control head in manipulating the computation tape in a Turing machine fashion or else may be used to place selected components in the constructing area and move welded blocks of components around. Arbib (1969, Chapter 10) shows that an instruction code can be specified for our basic modules so that it is not only possible to embed arbitrary Turing machines in the array of those components as in Figure 7.7, but also to program these machines so that they can construct other such machines in the construction area and to go on from there to show that there exists a universal constructor made of these components. It is then a standard procedure, following von Neumann's argument, to present an actual self-reproducing machine.

A CT-machine may be completely specified at any stage by a quadruple:

P. its program;
I. the instruction of the program it is executing;
T. the state of the tape (finite support);
C. the state of the construction area (finite support).

The construction we have just outlined yields

1 Theorem. *Any (P, I, T, C) configuration may be effectively embedded in the tessellation.*

2 Theorem. *There is an effective procedure whereby one can find, for a given CT-automaton A, an embedded CT-automaton $c(A)$ ("constructor of A") that once started, will proceed to construct a copy of A in the three rows of its constructing area immediately above it, and then activate that copy of A.*

Now, my procedure takes perhaps three cells in the program of $c(A)$ to code one cell of A, so it may still seem that any machine is only capable of constructing simpler machines.

We can effectively enumerate all the (P, I, T, C) configurations

$$M_0, M_1, \ldots, M_n \ldots$$

and then it takes a simple "Universal Turing Machine" type argument to prove

3 Theorem. *There exists a universal constructor M_u with the property that, given the number n, coded in binary form on its tape as I_n, it will construct the nth configuration M_n in its configuration area. Symbolically $I_n: M_u \to M_n$.*

Why isn't M_u self-reproducing? Because $I_u: M_u \to M_u$, but in the "second generation," $M_u \to$?, and reproduction fails. (A cell that produces a copy of itself minus the genes is not self-reproducing.)

4 Theorem. *For each recursive function h, there exists a machine M_c (where c depends on h) such that*

$$I_n: M_c \to M_{h(n)}.$$

PROOF. There clearly exists a program $P(h)$ of tape instructions that will convert I_n to the tape expression $I_{h(n)}$. The machine M_c then has for program $P(h)$ followed by the instructions of the program of M_u.   □

5 Theorem (Myhill, 1964). *For any computable function g, there exists a machine M_a such that*

$$M_a \to M_{g(a)}.$$

PROOF. Let $M_{s(x)} = I_x: M_x$. Then s is a computable function, and so is $g \circ s$. Thus, taking $h = g \circ s$ in Theorem 4, we have that there is a c such that

$$I_n: M_c \to M_{h(n)} = M_{g(s(n))}.$$

Setting $n = c$, we obtain

$$M_{s(c)} = I_c: M_c \to M_{g(s(c))},$$

and thus $a = s(c)$ satisfies the theorem.   □

For instance, g could be the function taking M into its mirror image, and so on.

6 Corollary. *There exists a self-reproducing machine.*

PROOF. Let $g(x) = x$ in Theorem 5. For the corresponding a, $M_a \to M_a$.   □

7 Theorem (Myhill, 1964). *Let $h(x, y)$ be a total computable function of two arguments. There is then a machine M_d for which, no matter what a, we always have*

$$I_a: M_d \to M_{h(a, d)}.$$

PROOF. Let us use $t_2(a, b)$ to denote the index of $I_a: M_b$, so that

$$M_{t_2(a, b)} = I_a: M_b.$$

By reasoning similar to the above, we may find r so that M_r, when given the two (appropriately punctuated) instruction tapes I_a and I_n, first decodes n to compute $t_2(n, n)$, then decodes a to compute $h(a, t_2(n, n))$, and then calls M_u to construct $M_{h(a, t_2(n, n))}$:

$$[I_a/I_n]: M_r \to M_{h(a, t_2(n, n))}.$$

Now set $d = t_2(r, r)$ and we have $M_{t_2(r, r)} = I_r: M_r$ so that

$$I_a: M_d = [I_a/I_r]: M_r \to M_{h(a, t_2(r, r))} = M_{h(a, d)}.$$ □

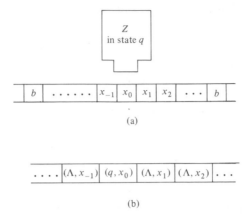

(a)

(b)

Figure 7.8   (a) A Turing machine Z. (b) A one-dimensional cellular space $\tilde{Z}$ which simulates Z in real time.

In Chapter 8, we shall present an amazing corollary: that there exists a machine whose descendants are always "smarter" (in the sense of being able to prove more theorems) than their parent.

Smith (1968) has shown that Lee's results on self-description of Turing machines (Theorem 6.2.3) may be used to construct self-reproducing configurations in a manner far simpler than that just outlined. We first note that for any Turing machine Z was can define a one-dimensional cellular space $\tilde{Z}$ that simulates Z in real time, with the state of each cell encoding a state-symbol pair of Z. Let Λ be a new "null state." Then the state space of $\tilde{Z}$ is just $(Q + \{\Lambda\}) \times X$, with (Λ, b) the quiescent state of the cellular automaton. As shown in Figure 7.8, the space $\tilde{Z}$ is just an "image" of the tape, with the square bearing symbol x_0 and scanned by the control box in state q being encoded by $\tilde{Z}$ state (q, x_0), while every other square with symbol x is encoded by (Λ, x). It is a straightforward exercise for the reader to write out the nearest-neighbor interaction rules for $\tilde{Z}$ that mimic the behavior of Z. For such a space, we say that Z is wired into $\tilde{Z}$, or that $\tilde{Z}$ has Z wired in.

Let the wired-in computer be a universal Turing machine U. We use the notation $x \underset{P}{\rightarrow} y$ to mean that Turing machine program P acting on initial tape x halts with y on the tape as its final result, so that we have

$$f(P), x \underset{U}{\rightarrow} y,$$

where f is the encoding function that encodes program P and tape x as the tape $f(P), x$ for U, and $x \underset{P}{\rightarrow} y$.

8 Lemma (Smith, 1968; see Lee, 1963). *For an arbitrary one-to-one recursive function f from (program, tape) pairs to tapes and for an arbitrary total recursive function g such that $g(x) = y$, there exists a machine with program P such that*

$$\underset{\uparrow}{x} \underset{P}{\rightarrow} f(P), x, y, \underset{\uparrow}{f(P), x.}$$

PROOF. Define the function h from programs to programs such that $h(Q)$ is a program that reads arbitrary input tape x, encodes program Q and tape x to get $f(Q)$, x, prints $f(Q)$, x, computes $g(x)$ to get y, prints y, prints $f(Q)$, x again (either by reencoding Q and x or by copying the result of the first encoding), and finally moves the head to the leftmost symbol in the rightmost encoding $f(Q)$, x. This is,

$$x \underset{\uparrow \, h(Q)}{\rightarrow} f(Q), x, y, f(Q), x.$$

Clearly h can be chosen total recursive. Thus, by the recursion theorem, there exists P that is a computational fixed point of h such that

$$x \underset{\uparrow \, P}{\rightarrow} f(P), x, y, \underset{\uparrow}{f(P)}, x. \qquad \square$$

Thus, in cellular space $\tilde{U}$ with U wired in, the following situation can hold:

$$f(P), x \underset{U}{\rightarrow} f(P), x, y, f(P), x$$

at some time $T > 0$. If we decree that V is U so modified that on completing such a computation it will backtrack to the rightmost comma, and start anew without changing anything to the left of that comma, we will obtain

$$f(P), x \underset{V}{\rightarrow} f(P), x, y, f(P), x, y, f(P), x$$

at some later time $T' > T$ and so forth. Hence, we have shown the following:

9 Theorem. *Let $\tilde{V}$ be a computation-universal cellular space with a modified universal Turing machine V wired in. Then there exists a configuration $c = f(P)$ in $\tilde{V}$ that is both self-reproducing and computes any given total recursive function g on supplied data x.* $\qquad \square$

This result puts us on guard against too hasty acceptance of the identity between a biological reality and any mathematical formalization. We shall pay somewhat more attention to an "automaton-theoretic biology" in Section 7.3.

7.2 Tessellations and the Garden of Eden

In Section 7.1, we saw examples of *tessellation automata* (or *cellular automata*) in which we have one finite automaton at each point, indexed by a pair of integers (i, j) of the plane. All the automata have identical state sets and transition functions. Each automaton receives as input the state of cells in a certain neighborhood. We single out a *quiescent state* q_0 such that if at time t an automaton and all its neighbors are in state q_0, then at time $t + 1$, the automaton will still be in state q_0.

Let G be the "cellular space," in this case the set of all pairs (i, j) of integers. Let Q be the set of states of the basic finite automaton. Then the state of all the cells at time t constitute a *configuration*, i.e., a map $c: G \to Q$ assigning a state to each cell of the tessellation, $c(g)$ being the state of the cell at g. We require that the support of c, $\sup(c) = \{g \mid c(g) \neq q_0\}$ be finite.

We have a global transition function F such that $F(c)(g)$ is the state of the cell at g at time $t + 1$, given that the configuration was c at time t. F has the property that if $\sup(c)$ is finite, then so too is $\sup(F(c))$.

A configuration c' is a subconfiguration of c if

$$c \mid \sup(c') = c' \mid \sup(c'),$$

that is,

$$c'(g) \neq q_0 \Rightarrow c'(g) = c(g).$$

c is called *passive* if $F(c) = c$, that is, it does not change with time; and it is called *completely passive* if every subconfiguration of c is passive. If W is the set of states that occur in the configuration c, we say that W is the *alphabet* of c, and c is called a *configuration over* W.

A subset W of Q is called *passive* or *completely passive* if all configurations over W are such.

We say that configurations c and c' are *disjoint* if their supports are. If c and c' are disjoint, we define their *union* by

$$(c \cup c')(g) = \begin{cases} c(g) & \text{if } g \in \sup(c), \\ c'(g) & \text{if } g \in \sup(c'), \\ q_0 & \text{otherwise.} \end{cases}$$

The reader should realize that the union of two completely passive configuration is not *necessarily* completely passive.

One usually studies computation in cellular arrays by "embedding" Turing machines in the tessellation. We have followed this procedure here but we want to emphasize that this is highly unsatisfactory, and does not take into account the full parallel processing power of the array. To move away from this to an exploration of the real subtleties and possibilities of parallel computation on a modular computer (i.e., tessellation-embedded computer) is a vital challenge at the interface of automata theory with the design of real computers, and we have seen something of developments in this area in earlier chapters. Such efforts do not leave the domain of Turing-computable (i.e., recursive) functions, but do yield structures that compute the functions far more efficiently than with a Turing machine, or even with a collection of multihead Turing machines.

We close this section by presenting results from E.F. Moore's (1962) "Machine Models of Self-Reproduction."

A configuration c *contains n copies* of a configuration c, if there exists n disjoint subsets of the array of c and each of these subsets is a copy of c. A configuration c will be said to be *capable of reproducing n offspring* by time T if starting with a copy of c with the remaining cells quiescent at time 0, there is a time $T' > T$ at which the set of all nonquiescent cells will be an array whose configuration contains at least n copies of c.

A configuration is *self-reproducing* in the sense of Moore if for each positive integer n, there exists T such that c is capable of reproducing n offspring by time T.

1 Theorem. *If a self-reproducing configuration is capable of reproducing $f(T)$ offspring by time T, then there exists a positive real number k such that $f(T) \leq kT^2$.*

PROOF. Let c be the self-reproducing configuration. Let the smallest square array large enough for a configuration containing a copy of c be of size $D \times D$. Then at each time T, the total number of nonquiescent cells is at most $(2T + D)^2$. If r is the number of cells in the array of c, then $f(T) \leq (2(T-1) + D)^2/r$.   □

By an *environment* is meant a specification of states for all cells of the tesselation except those of a square piece. By the insertion $E(c)$ of a configuration c (some cells of which may be quiescent) of appropriate size into an environment E is meant simply the result of specifying the states of the unspecified cells of E to be the states of the corresponding cells of c.

Two configurations c_1, c_2, of the same size are said to be *distinguished* by the environment E if $F(E(c_1)) \neq F(E(c_2))$.

2 Lemma. *Let E_0 be the environment consisting entirely of passive cells. If every pair of distinct configurations can be distinguished by some environment (possibly depending on the choice of configurations), then for any two distinct configurations c_1, c_2 we have $E_0(c_1) \neq E_0(c_2)$.*

PROOF. If $E_0(c_1) = E_0(c_2)$, and c_1^* and c_2^* are obtained by adjoining to c_1 and c_2 a border of passive cells of width 2, then c_1^* and c_2^* would have identical sequents in *every* environment.   □

c is called a *Garden-of-Eden* configuration if there is no c' such that c is a subconfiguration of $F(c')$; that is, c can only be set up in the tessellation by an external agency.

3 Lemma. *Given $A > 1$ and $n > 1$, there exists a positive integer k such that*

$$(A^{n^2} - 1) < A^{(kn-2)^2}.$$

PROOF. Since $A^{n^2}/(A^{n^2} - 1) > 1$, we may choose k so large that

$$\log_A\left(\frac{A^{n^2}}{A^{n^2} - 1}\right) > \frac{4n}{k} - \frac{4}{k^2} > 0$$

then

$$\frac{A^{n^2}}{A^{n^2} - 1} > A^{(4n/k - 4/k^2)}$$

$$A^{n^2 - (4n/k) + (4/k^2)} > A^{n^2} - 1$$

$$A^{n^2} - 1 < A^{[n - (2/k)]^2};$$

that is,

$$(A^{n^2} - 1)^{k^2} < A^{(kn - 2)^2}. \qquad\qquad \square$$

4 Theorem. *A tessellation has a Garden-of-Eden configuration iff there are two distinct but indistinguishable configurations.*

PROOF. Fix n and choose k as in Lemma 3.

Suppose there is an $n \times n$ Garden-of-Eden configuration G. A $kn \times kn$ square is made up of k^2 $n \times n$ squares. There are A^{n^2} configurations for an $n \times n$ array, but at least one of these, G, is Garden-of-Eden. Thus the number of non-Garden-of-Eden $kn \times kn$ arrays is $\leq (A^{n^2} - 1)^{k^2}$.

Suppose every distinct pair of configurations is distinguishable. Choosing k as in Lemma 3, we see, in particular, that distinct $(kn - 2) \times (kn - 2)$ configurations, embedding in the blank configuration, must have distinct sequents. But the nonblank portion of such a sequent is contained in a $(kn \times kn)$ square. Thus the number of distinct non-Garden-of-Eden $(kn \times kn)$ configurations is at least equal to the total number of distinct $(kn - 2) \times (kn - 2)$ configurations. But this implies that $A^{(kn-2)^2} \leq (A^{n^2} - 1)^{k^2}$, contradicting Lemma 3. Thus either at least two distinct $(kn - 2) \times (kn - 2)$ configurations are indistinguishable, or there is no Garden-of-Eden configuration. $\qquad\qquad \square$

7.3 Toward Biological Models

We wish to show how to modify the tessellation model so that it may provide a logical framework for understanding embryological processes. The great excitement over DNA–RNA prompted many people to claim that at last we had found the "secret of life." However, unravelling the DNA $\rightarrow$ RNA and RNA $\rightarrow$ enzyme transductions, while important biologically, are of little interest in the present context because they do not address the challenging question "How can a complex multicellular automaton grow from a single cell, given that a finite program can be executed within each cell?" Such a question as this is nontrivial, and becomes a fit topic for automata theory, although it must be confessed that, at present, the emphasis is on ingenious programming of cellular arrays rather than on weaving a rich texture of theorems. We saw how a multicellular organism could construct a copy of itself, but we did not answer the question "How can a complex multicellular automaton grow from a *single* cell." In this section, I want to sketch how our model can be modified to yield an answer, and wherein that model differs from the embryological situation.

Let us first note that the self-reproduction we have studied is a far narrower concept than the embryological one—a human zygote grows into a human, not into a replica of one of the parents. The zygote contains only an outline, a program that, in interaction with the environment, produces an organism of a certain species (we are now discarding questions of mutation and evolution). This leads to a whole series of questions—which here I can merely raise—of the "identity" of an automaton. What does it mean to say two automata belong to the same "species"? Embryological reproduction gives rise to off-spring with similar structure, and this implies similarity of function. What are measures of structural similarity and functional similarity, such that the first implies the second? Are there interesting classes of automata for which we may carry out decompositions into a species-dependent automaton and an individuality-expressing automaton where the decompositions are at the functional, rather than structural, level? Can we study reliability of reproduction for such systems, where the genetic information determines the species-dependent automaton with high probability, random influences making their appearance chiefly in modifying the individuality-expressing automaton?

In Section 7.1, we used modules (Figure 7.6) far more complex than von Neumann's 29-state elements. This seems admirable, rather than sad, now that we are turning to biological questions (in fact, we shall shortly introduce a model with far larger modules). The living cell, with its synthetic machinery involving hundreds of metabolic pathways, can rival any operation of our module, as well as being under the control of DNA molecules with far more bits of information than our cell can store. So, perhaps, we lose biological significance by unduly limiting the information content of the module.

Our construction rests on the assumption that we can produce new cells at will, our only problem being to ensure that they contain the proper instructions. Given cell reproduction, how do we replicate an organism? This is the topic we treat—and it makes sense to ignore a lot of hard work by using complicated cells. In fact, we might hypothesize that "sophisticated" organisms can evolve (by whatever mechanism) only when there are complicated reproducing cells available. Contrasting our model with organism reproduction, we note that

(1) Our program was embedded in a string of cells, whereas the biological program is a string stored in *each* cell.
(2) We use a complete specification, whereas "Nature" uses an incomplete specification.
(3) We did not use anything like the full power of our model (i.e., the operation was sequential instead of parallel).
(4) We constructed a passive configuration—we set up all the cells with their internal program, and only then did we activate the machine by telling the

control head to execute its first instruction. Contrast the living, growing embryo. Our construction *relied* on the passivity of the components, and demanded that any subassembly would stay fixed and inactive until the whole structure was complete. The biological development depends on active interaction and induction between subassemblies.

The logic of taking factors (2)–(4) into account is hard, and we do not treat it here. To modify our model to take account of factor (1), we still think of a cell of the tessellation as corresponding to a cell of the organism—with the active cells in the construction area corresponding to the embryo—but now each cell contains the whole program. So in a sense activation of different subroutines in a cell of our modified model would correspond to differentiation of cells of the embryo. Each model stores a whole program, but with only a substring loaded into the 22 instruction registers. Biologically, we put all the available information in one cell and let it grow. In our new model, we can think of the machine as starting as a single cell with two strings (one corresponding to the Mark I program of Figure 7.6, one to the tape). It secretes new cells and manipulates tape to produce a discrete aggregate of cells. Starting with one cell containing this information in our model, we would program it to secrete extra cells for tape as well as for the growing organism. The tape cells are discarded "at birth." (We still have not used parallel processing—for we put a large but bounded amount of information in the individual cells, and then used extra cells for tape—this captures nothing of the embryological organization. "A germ cell doth not a living embryo make.")

We have not said anything about where cells came from: this corresponds to the question of the evolution of life. There are two distinct problems of evolution. One is, starting from a relatively unstructured universe, how do you get cells? The next evolutionary problem is how do cells start aggregating? I think at the present moment it is relatively easy, at least qualitatively, to get the idea of cells competing for various nutrients in the environment, cooperating to form aggregates, and these aggregates then evolving in a classical domineering fashion. The question of where the cells came from is a very different and very difficult one.

The questions of how reproducing cells evolved in the first place is outside the scope of the present volume, but should be borne in mind. Codd (1965) considers tessellations with even simpler components than von Neumann's. A pure automata problem is to embed our module in Codd's model, where one of our cells is simulated as an aggregate of Codd's cells with appropriate change of time scale. Perhaps we can approach the cellular-evolution problem by imagining a subtessellation with components comparable to the macromolecules of biology, and consider reproduction of our modules as aggregates of these pseudomacromolecules. Our constructions would then treat arrays of arrays.

The Mark II module is shown in Figure 7.9. We have kept the tessellation structure, side-stepping the morphogenesis of individual cells. The control

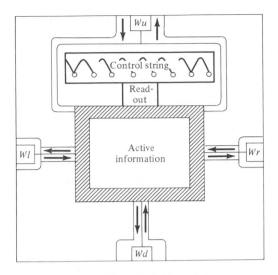

Figure 7.9   Mark II module.

string is segmented in words that correspond to the possible instructions of the original modules. The whole control string corresponds to the whole CT program in our original model. Only a small portion of the control string can be read by an individual cell. *Every cell in an organism has the same control string.* Individual cells differ only in the portion of the control string that can currently be read out. *The change in activation of portions of the control strings is our analog of differentiation. The increase in the number of cells in a co-moving set is our analog of growth.*

A cell computes under the control of its activated control-string portion and of its inputs. Thus a cell that functions one way as part of a co-moving set may well behave in a completely different way when removed from the aggregate—a phenomenon familiar in embryological experiments.

Besides computation on inputs to produce outputs, a cell can divide (this corresponds to the activity in the construction site in the original model) or die (this corresponds to returning a cell to the quiescent state).* Thus we are really considering the quiescent cell of Mark I as a noncell in the Mark II model, and our tessellation is to be thought of as empty save where there are moving sets. These masses grow only when individual cells divide. The "active information" region corresponds to the bit register and 22 instruction registers of the Mark I module. It has access to the activated portion of the control string.

When a cell divides, the original cell may be considered as preserved at its original site, while a replica has been produced at a neighboring site. This replica will have the same control string as the original cell, but may be

---

* To obtain sexuality one "simply" requires a means for two cells to redistribute the instructions of their control strings.

"differentiated" in that the activated region of the control string may be adjacent to, rather than identical with, that of the original cell.

Given an automaton A in the Mark I model, we may replace it by a Mark II automaton simply by replacing each cell by a Mark II cell. We then grow A from a single cell by encapsulating the program of the constructor within a single cell $\hat{c}(A)$. This single cell grows into the Mark II version of $\hat{c}(A)$ simply by "reading out" the control string into a linear string of cells, the last two (leftmost) of which then produce a control head. This proceeds to build a copy of A.

The problem of self-reproduction becomes very simple in this model. We simply require that $\hat{c}(A)$ initially produces a copy of itself (a single cell!) that is rendered dormant, $\hat{c}(A)$ then produces $c(A)$, $c(A)$ in turn builds A but before turning A loose, attaches to it the "germ cell," i.e., the copy of $\hat{c}(A)$. A then reproduces by releasing this copy of $\hat{c}(A)$ into the tessellation at "maturity."

It is worth noting that, in this model, the control string is not that of the reproducing automaton $\hat{c}(A)$ itself, but rather that of the constructor of A. At this stage, I would not push this observation as a clue to the operation of real biological systems. Rather, its purpose is to shock the biologist, and the theoretician, into a reappraisal of how devious genetic coding may be. Our progress in decoding the fashion in which DNA directs the production of proteins has been so impressive that we tend to forget how little we know about the way this relates to the morphogenesis of the multicellular organism. I should also emphasize that we have considered the basic units of our control string as explicit instructions with respect to cellular computation or reproduction. This must be contrasted with the exceedingly indirect commands in the control string (DNA) of a real cell, which serve to bias overall metabolic activity rather than explicitly modify specific bits of data storage.

Let me reiterate that this model shares with the Mark I model an essentially serial mode of operation. We shall have a far greater feel for the demands we must make on a genetic coding system when we have programmed the Mark II system in a parallel growth mode, in which *every cell of an organism may be considered a growth point.*

In "designing" an organism, we must use subroutines to build "components," i.e., tissues with a basically repetitive structure. Higher-level routines then serve to arrange these components in appropriate topographical relations. This is one way in which we reduce the length of the program required to specify the growth of an organism. Another factor that helps economize the growth program is that *an organism will only grow normally in an appropriate highly structured environment.* Thus the growth program may rely on "self-organizing mechanisms" (cf. Section 4.3) to select information out of the environment, thus economizing on the information required in the zygote. That is why both nature and nurture are important.

The reader who wishes to study further the "classical" theory of cellular automata can do no better than consult the collection edited by Arthur Burks

(1970). Cellular automata have once again become fashionable, and the "state of play" is briefly surveyed by Hayes (1984). On the other hand, readers who are more drawn by formal models of embryological processes should turn to Ransom (1981) for a survey of a variety of approaches, in addition to those based on cellular automata.

References for Chapter 7

Much of this chapter was adapted from Chapter 10 of Arbib (1969), while Section 7.3 is based on Sections 4 and 5 of Arbib (1967).

Arbib, M.A., 1966, A simple self-reproducing universal automaton, *Inform. Control* **9**: 177–189.

Arbib, M.A., 1967, Automata theory and development: Part I, *J. Theoret. Biol.* **14**: 131–156

Arbib, M.A., 1969, *Theories of Abstract Automata*, Prentice-Hall.

Burks, A.W., Ed., 1970, *Essays on Cellular Automata*, University of Illinois Press.

Codd, E.F., 1965, Propagation, computation and construction in 2-dimensional cellular spaces, Technical Publication of the University of Michigan. Reprinted in 1968 as *Cellular Automata*, Academic Press.

Hayes, B., 1984. Computer recreations: The cellular automaton offers a model of the world and a world unto itself, *Scientific American*, March, **250**, 3: 12–21.

Lee, C.H., 1963, A Turing machine which prints its own code script, in *Proc. Symp. Math. Theory of Automata*, Polytechnic Press, pp.155–164 (Vol. 12 of the Microwave Research Institute Symposia Series).

Moore, E.F., 1962, Machine models of self-reproduction, *Math. Prob. Biol. Sci., Proc. Symp. Appl. Math.* **14**: 17–33.

Myhill, J., 1964, The abstract theory of self-reproduction, in *Views on General Systems Theory*, (M.D. Mesarovic, Ed.), John Wiley and Sons, pp. 106–118.

Ransom, R., 1981, *Computers and Embryos: Models in Developmental Biology*, John Wiley and Sons.

Smith, A.R., 1968, Simple computation-universal cellular spaces and self-reproduction, *Conf. Record, IEEE 9th Ann. Symp. Switching and Automata Theory*, pp. 269–277.

Thatcher, J.W., 1970, Universality in the von Neumann cellular model, in *Essays on Cellular Automata* (A.W. Burks, Ed.), University of Illinois Press, pp. 132–186.

Turing, A.M. (1936) "On Computable Numbers, with an Application to the Entscheidungs-Problem" *Proc. London Math. Soc.*, Ser. 2–24, 230–65, with a correction, *Ibid.*, Ser. 2–43 (1936–7), 544–46.

von Neumann, J., 1951, The general and logical theory of automata, in *Cerebral Mechanisms in Behavior: The Hixon Symposium*, John Wiley and Sons.

von Neumann, J., 1966, *Theory of Self-Reproducing Automata* (edited and completed by A.W. Burks), University of Illinois Press.

CHAPTER 8
Gödel's Incompleteness Theorem

In this, our final chapter, we shift our center of interest first to the foundations of mathematics. In Section 8.1, we shall give a brief historical review of the formalist approach to the foundations of mathematics and see how Gödel's incompleteness theorem invalidated much of the Formalist program. In Section 8.2, we shall discuss some general properties of recursive logics, yielding a proof of Gödel's Incompleteness Theorem. We shall also follow Myhill's surprising result that we can effectively remove this incompleteness, although never totally but only a part at a time. To deepen our understanding of Gödel's work, we present a proof of his Completeness Theorem in Section 8.3, and show the way in which completeness and incompleteness coexist. Section 8.4 studies speed-up theorems: showing that adding an axiom to a logic may not only enable the proofs of new theorems, but also dramatic shortening of proofs that were already available. Finally, in Section 8.5, we return to the main theme of this book by discussing the philosophical controversy centering around the implications of Gödel's theorem for the question: Are brains essentially superior to machines?

My desire to prove Gödel's Incompleteness Theorem is thus essentially threefold: first, because of its implications for the brain–machine controversy; second, because of its importance in the foundations of mathematics; and third, because the proof given in the book is sufficiently short and simple to help dispel the myth that the proof of Gödel's theorem is accessible only to the specialist in mathematical logic.

8.1 The Foundations of Mathematics

The philosopher Kant proclaimed that the axioms of Euclidean geometry were given *a priori* to human intuition. This proclamation was in the spirit of the definition of an axiom of a logical system such as Euclidean geometry as a self-evident truth, a definition that had stood unquestioned for 2,000 years.

This whole attitude received a severe jolt in the 19th century from the work of Bolyai, Lobachewsky, and Riemann. They postulated systems of geometry that were non-Euclidean—more specifically, systems that denied the truth of the following *axiom* of the Euclidean scheme: "Given a line, and a point not on it, then there exists precisely one line through the given point parallel to the given line." And we now live in an age in which the accepted view of the universe afforded us by the Einsteinian theory of relativity involves a space whose geometry is best described by a Riemannian non-Euclidean scheme in which parallel lines just do not exist.

In other words, our present view not only contradicts Kant's view that the Euclidean axioms are given *a priori* to human intuition, it even asserts that one of Euclid's axioms, the so-called parallel axiom, is actually untrue as a description of the universe. We now believe that Euclidean geometry is quite accurate enough to describe the spatial relations of our everyday lives, but not the spatial entirety of our universe. As a result, however, we are now faced with a genuine and important question which the Kantian viewpoint allowed us to dispose of effortlessly: How do we know that Euclidean geometry is consistent (i.e., free from contradictions)?

Just to aggravate the enormity of this question, *Riemann showed that if Euclidean geometry was consistent, then his non-Euclidean geometry was consistent*. In fact, his model of non-Euclidean geometry interpreted "points" as "pairs of diametrically opposite points on a sphere," and "lines" as great circles on the sphere. Clearly, all "lines" meet, so the parallel axiom does not hold, but all the other Euclidean axioms do hold—as a result of holding for a 3-dimensional Euclidean space in which the sphere may be embedded. Here was a fine contretemps—not only had the consistency of the Euclidean scheme ceased to be *a priori* evident, but it was shown that its consistency implied that of a rival scheme! One evident conclusion was that the students of axiomatic systems had to tackle the problem of consistency quite independently of any questions as to whether or not the system afforded an apparently "true" description of the "real world."

Meanwhile, developments in set theory were showing that the consistency of a system could not be checked by mere common sense. In fact, the system of set theory propounded by Cantor seemed completely consistent until Russell, among others, pointed out that this apparently "safe" system contained an annihilating paradox, which runs as follows: Consider the set of mathematicians—it is not a mathematician, and so this set does not belong to itself. However, the set of things talked about in this chapter is talked about in this chapter, so this set does belong to itself. Hence, we may define N to be the set of all those sets that do not belong to themselves. Thus

M belongs to N   if and only if   M does not belong to M.

So the set of mathematicians belongs to N, but the set of things described in this chapter does not. Does N belong to N? According to the above

N belongs to N   if and only if   N does not belong to N.

A paradox! And so naive set theory is inconsistent. Now Russell avoided such contradictions by introducing his Theory of Types, but the point here is that consistency is not an evident property of a logical system.

Riemann had shown that his geometry is consistent if Euclidean geometry is consistent, and Hilbert showed that Euclidean geometry is consistent if arithmetic—the elementary theory of the positive whole numbers—is consistent. The problem was thus to exhibit a consistent axiomatic system of arithmetic.

Now, a logical system has a collection of axioms from which we obtain theorems by the repeated application of a number of rules of inference. The school of Formalists, led by Hilbert, decided that in their search for proofs of consistency they would ignore all questions of the truth and meaning of the axioms and theorems and would instead regard the axioms as mere strings of symbols and the rules of inference only as methods of obtaining new strings. They further decided to require that the rules of inference operate in a purely finite well-determined manner, such as is exemplified very well in the operation of the Turing machines we studied in Chapter 6. Let us then agree to call a logical system satisfying such conditions a *recursive logic*. If we wish to use a recursive logic to describe the theory of positive integers, we must equip it with symbols that correspond to the basic quantitative notions of elementary number theory.

The Formalists were searching for a consistent arithmetical logic that was *complete*, i.e., in which one could prove as theorems all true statements about the integers. Furthermore, they demanded that the consistency of the system be shown in a manner as safe, well determined, and "finitistic" as that in which the rules of inference were to operate.

This Formalist program was wrecked by Gödel's Incompleteness Theorem, first expounded in his famous paper (1931) on formally undecidable theorems of *Principia Mathematica* and related systems.

His theorem states that *any* adequate consistent arithmetical logic is incomplete, i.e., there exist true statements about the integers that cannot be proved within such a logic. This important result (which we shall prove in the sequel) showed that the Formalist search for a complete consistent arithmetical logic was doomed to failure. Acutally, Gödel showed even more—namely, that it was impossible to show that an arithmetical logic (admittedly incomplete) was consistent by methods *that could be represented in the logic itself*.

Gentzen has since proved elementary number theory consistent but by using "ε_0-induction," which is an infinite extension of the familiar technique of mathematical induction—a method not satisfactory to the Formalists because it is not "finitistic." The present situation as regards the search for a consistency proof for arithmetic is thus as follows: Gödel has shown that there exists no finitistic proof expressible within arithmetic itself; Gentzen has given a proof, but only by using methods whose consistency is perhaps as open to doubt as that of the system they were called in to justify; and the question of whether or not there exists a consistency proof not expressible within arithmetic but nevertheless "finitistic" is still open.

8.2 Incompleteness and Its Incremental Removal

In Theorem 6.3.9, we saw that there exist recursively enumerable sets that are not recursive, i.e., a machine can generate the elements of such a set, but there is no machine that can effectively test whether or not an arbitrary given number belongs to the set. The reader may be interested to learn that this result is an abstract form of Gödel's celebrated *incompleteness theorem*, which states that any axiomatic system whose vocabulary is *adequate* to express a sufficiently rich collection of statements about numbers, and which is *consistent* (i.e., its axioms do not imply any contradictions), must also be *incomplete* in that there are truths about numbers expressible, but not provable, in this axiomatic theory. Of course, the tough thing in studying a given axiomatic system is to show that it is consistent and adequate. We give an informal proof of Gödel's theorem.

A *recursive logic* is specified by a finite set of *axioms* and a finite set of *rules of inference*—effective procedures for telling whether or not a statement may be deduced from other statements. A *proof* is a finite sequence of statements each of which is either an axiom, or deducible from earlier statements in the sequence by one of the rules of inference. A *theorem* is any statement that may occur in a proof.

1 Observation. *The proofs of a recursive logic form a recursive set, while the theorems form a recursively enumerable set.*

PROOF.

(i) We can effectively check whether a sequence of statements is a proof by (a) checking each line to see if it is an axiom or, then, (b) checking the line against each subset of earlier lines to see if it is deducible from them by one of the rules of inference.

(ii) Pick some effective enumeration $\{p_0, p_1, \ldots, p_m, \ldots\}$ of the proofs. Then we may obtain the set of theorems as the range of the total computable function that, given n, generates the last line of the proof p_n. $\square$

Adequacy means that for each recursively enumerable set U and for each $n \in N$ there is a string $U(n)$ which is to represent the statement $n \in U$—and we require that, given any string, we can tell if it is of the form $U(n)$, and if so, for what U and what n; and that $U(n)$ is a theorem if and only if $n \in U$. *Consistency* means that for no w can both w and its negation $\neg w$ be theorems. We shall use $\Sigma \vdash w$ as shorthand for "w is a theorem of the logic Σ". Hence consistency and adequacy entail that $n \notin U$ whenever $\Sigma \vdash \neg U(n)$.

2 Gödel's Incompleteness Theorem. *Every consistent adequate arithmetical logic is incomplete.*

PROOF OUTLINE: Let, now, U be a recursively enumerable set that is not recursive. We show that the system is *incomplete* by showing that there exists an n for which neither $U(n)$ nor $\neg U(n)$ is a theorem. Since either $n \in U$ or $n \notin U$, one of these represents a true statement, and so must be provable if the system is to be complete.

Suppose, by way of contradiction, that for each n one of $U(n)$ and $\neg U(n)$ is a theorem. Then given n we have an effective procedure for telling whether or not n belongs to U: Generate, one by one, the theorems (remember, they form a recursively enumerable set)—eventually either $U(n)$ or $\neg U(n)$ will be encountered. If it is $U(n)$, we decide that $n \in U$; if it is $\neg U(n)$, we decide that $n \notin U$. Thus if the system were complete, U would have to be recursive, which it is not. We conclude that the system is incomplete. $\square$

Recalling from Chapter 6 our notation S_n for the nth set in our effective enumeration of the recursively enumerable sets, we can in fact obtain even more information by the above type of reasoning. First we explore in more detail the possible anatomy of nonrecursive sets.

3 Definition. A set R is *creative* if it is recursively enumerable and if there exists a total recursive function f such that

$$S_n \subset \bar{R} \Rightarrow f(n) \in \bar{R} - S_n.$$

Thus, no creative set is recursive, for if it were, its complement $\bar{R}$ would be recursively enumerable. However, $\bar{R}$ cannot, by the above definition, equal S_n for any n.

4 Fact. $K = \{n \mid n \in S_n\}$ *is creative.*

PROOF. Since $\bar{K} = \{n \mid n \notin S_n\}$, $S_n \subset \bar{K}$ implies $n \in \bar{K} - S_n$, and so K is creative with $f(n) = n$. $\square$

5 Fact. *If a set R is creative, then its complement contains an infinite recursively enumerable subset.*

PROOF. Let f be the function corresponding to R in Definition 3. We form a sequence $n_0, n_1, \ldots, n_m, \ldots$ such that $f(n_i) \neq f(n_j)$ whenever $i \neq j$, yet every $f(n_i)$ lies in $\bar{R}$.

Recall that S_n is the set $\{\phi_n(m) \mid m \in \mathbf{N}\}$ of results generated by the Turing machine Z_n in response to inputs from $\mathbf{N}$. We define the total recursive function $h: \mathbf{N} \to \mathbf{N}$ as follows. Given n, we compute $f(n)$, then modify Z_n so that it returns $f(n)$ for input 0, but returns $\phi_n(m)$ for input $m + 1$. Let $Z_{h(n)}$ be the resulting machine; i.e., $h(n)$ is the encoding of the program we have just described. Then it is clear that for all n in $\mathbf{N}$,

$$S_{h(n)} = f(n) \cup S_n.$$

We then effectively enumerate a set $\{n_0, n_1, n_2, \ldots, \}$ as follows. Let n_0 be an index for the empty set, $S_{n_0} = \emptyset \subset \bar{R}$. Thus $f(n_0) \subset \bar{R} - S_{n_0}$, and so $S_{h(n_0)} = \{f(n_0)\} \cup S_{n_0} \subset \bar{R}$. By induction, we have that $n_m = h(n_{m-1})$ is such that $S_{n_m} = \{f(n_0), f(n_1), \ldots, f(n_{m-1})\}$ is a set of m *distinct* elements contained in $\bar{R}$. Thus $\{f(n_m)|m \geq 0\}$ is an infinite recursively enumerable subset of $\bar{R}$. $\quad\square$

However, not all recursively enumerable sets are such that their complements have infinite r.e. subsets.

6 Definition. A set S is *simple* if it is recursively enumerable while $\bar{S}$ is infinite but contains no infinite recursively enumerable subset. Thus, no simple set is recursive.

7 Fact. $S = \{x|\exists n \text{ such that } x \text{ is the first element of } S_n \text{ with } x > 2n\}$ *is simple.*

PROOF. If S_n is infinite, S_n certainly contains an element x that is greater than $2n$, and so $S_n \cap S \neq \emptyset$. Thus no infinite recursively enumerable set can be contained in $\bar{S}$. However, $\bar{S}$ is infinite because the x's in $\{0, 1, \ldots, 2n + 1\}$ may only gain entry to S through the sets $S_1, \ldots, S_n$, so that at least half of any set $\{0, 1, \ldots, 2n + 1\}$ belongs to $\bar{S}$. We conclude that S is simple. $\quad\square$

Now let us reassess the proof of Gödel's theorem. Implicit in it is that for any adequate consistent arithmetical logic Σ, $\{n|\Sigma \vdash \neg U(n)\}$, the set of n for which $\neg U(n)$ is a theorem of Σ, is a recursively enumerable subset of $\bar{U}$.

Thus if U is simple, $\{n|\Sigma \vdash \neg U(n)\}$, being a recursively enumerable subset of $\bar{U}$, is finite, even though $n \notin U$ for infinitely many n. Note well that this is true for *any* choice of an adequate consistent logic Σ.

If K is creative, then $S_m = \{n|\Sigma \vdash \neg K(n)\} \subset \bar{K}$, and so $f(m) \in \bar{K}$, where $f(m) \notin S_m$. If we now construct a new logic Σ' by adjoining $\neg K(f(m))$ as an additional axiom, we have a logic in which the truth $f(m) \in \bar{K}$ is represented by a theorem.

Now, for each logic Σ, we may specify a Turing machine $Z_{h(\Sigma)}$ that, started on a blank tape, proceeds to print out (scratchwork and) an effective enumeration of the theorems of Σ. Furthermore, the passage from Σ to $h(\Sigma)$ is effective, and given any n, we can tell whether it is an $h(\Sigma)$, and, if so, for which Σ.

Given a Turing machine Z_n, let $g(n) = n$ if Z_n is not $Z_{h(\Sigma)}$ for some logic Σ. If $n = h(\Sigma)$, let $Z_{k(n)}$ be Z_n modified so that it prints out only theorems of the form $\neg K(n)$ for our creative set K. But then $\neg K(k(n))$ may be consistently adjoined to the axioms of Σ to yield a new logic Σ'. Let $g(n) = h(\Sigma')$. Clearly, then, g is a total recursive function, and we have

8 Theorem (Myhill, 1964). *There is a total recursive function g such that for the Turing machine $Z_{h(\Sigma)}$ that prints out theorems of the adequate consistent arith-*

metical logic Σ, *the Turing machine* $Z_{g(h(\Sigma))}$ *is* $Z_{h(\Sigma')}$ *for an adequate consistent arithmetical logic* Σ' *with more theorems than* Σ.

Thus, while Gödel's incompleteness theorem points to an inevitable limitation of any axiomatization of arithmetic, Myhill's theorem points out the much less well known fact that this limitation can be *effectively* overcome. And, of course, the process may be iterated mechanically again and again. We conclude this section with an amusing application of this due to Myhill (1964) in the study of self-reproducing automata.

We recall from the abstract theory of constructors of Chapter 7 that Theorem 7.1.7 states that if $h(x, y)$ is a total recursive function of two arguments, there is then a machine M_d for which always

$$I_a: M_d \to M_{h(a,d)}.$$

Now let $M_a < M_b$ mean that M_b prints out all the strings that M_a prints and more, and all the strings that M_b prints our are true statements of arithmetic. We shall prove:

9 Theorem. *There exists an infinite sequence of machines* $\{M_{z_i}\}$ *such that we have simultaneously*

$$M_{z_i} < M_{z_{i+1}}$$

and

$$M_{z_i} \to M_{z_{i+1}}.$$

PROOF. Simply take $M_{h(a,d)} = I_{g(a)}: M_d$. □

We may call each M_{z_i} a machine each of whose descendants "outsmarts" its predecessor. Myhill observes that the theorem is a brutal parody of the growth of intelligence, but is of methodological significance in that it suggests the possibility of encoding a potentially infinite number of "directions to posterity" on a finitely long "chromosomal" tape.

8.3 Predicate Logic and Gödel's Completeness Theorem

In the last section, we traced the essential reason for the incompleteness of a recursive logic: If the logic were consistent and complete, it would yield a decision procedure for each represented set. Given n and U, enumerate the theorems of the logic. If $U(n)$ is listed, decide that $n \in U$. If $\neg U(n)$ is listed, decide that $n \notin U$. But we know that there are recursively enumerable sets that are not recursive . . . .

However, our presentation was very abstract. In this section, we look in more detail at how formulas can be written in predicate logic, prove another famous theorem of Gödel's, the *completeness* theorem, and then sketch the original proof that Gödel gave of his incompleteness theorem.

No, this is *not* a contradiction! The incompleteness theorem tells us that a logic is incomplete because there are true facts about arithmetic that cannot be proved as theorems of the logic. The logic is thus a *partial* description of arithmetic, and so may well be (in fact is) a partial description of many systems other than normal arithmetic. In the same way, we saw that Euclid's axioms, minus the parallel axiom, were not only satisfied by the points and lines of plane geometry, but also by the points and great circles of spherical geometry. We are going to make precise the notion that a *relational structure* (e.g., arithmetic, plane geometry) *satisfies* a logic. Incompleteness of a logic leaves room for many different models to satisfy its axioms and rules of inference. The completeness theorem asks us to contemplate *all* these models, and consider those statements that are true in *every* model. The question is: Are they all theorems, i.e., provable, in the logic? The answer is yes.

In summary, a logic that is incomplete with respect to the truths of arithmetic is nonetheless complete with respect to those truths that are held in common by all the models that satisfy the logic. Let us, then, turn to the formalization of all this.

Our first aim will be Definition 3, the definition of a wff (well-formed formula). This is akin to the notion of a grammatically correct sentence in English. Having obtained a core set of wff's, we build new ones both by propositional connectives and by quantifiers:

1 Propositional Connectives. Given wff's A and B in a logic, we wish to be able to combine them in various ways to obtain new wff's, such as:

$$\neg A \qquad \text{not } A$$

$$A \rightarrow B \qquad A \text{ implies } B$$

$$A \wedge B \qquad A \text{ and } B$$

$$A \vee B \qquad A \text{ or } B \text{ (or both)}$$

$$A \leftrightarrow B \qquad A \text{ if and only if } B$$

The models we consider are *two-valued*, i.e., any *statement* is to be interpreted as either true or false; while a *predicate* such as $x > y$ or Boy (x) or $f(x + 1) = x \cdot f(x)$ only assumes a truth value upon substitution of constants for all its variables. We require that every model interprets the *propositional connectives* given above in the same way as two-valued functions, e.g., $A \wedge B$ is true when A is true and B is true; and $A \wedge B$ is false otherwise.

We can express this in a "truth table," which tabulates the truth value of $A \wedge B$, given the truth values of A and of B. (For more on truth tables, see, e.g., M.A. Arbib, A.J. Kfoury, and R.N. Moll, *A Basis for Theoretical Computer Science*, Springer-Verlag, 1981.)

A	B	$A \wedge B$
T	T	T
T	F	F
F	T	F
F	F	F

We require $A \to B$ to be true whenever both A is true and B is true. We also require that if A is true and $A \to B$ is true, then we can infer that B is true. One scheme of assigning truth values to $A \to B$ consistent with these requirements is as follows:

A	B	$A \to B$
T	T	T
T	F	F
F	T	T
F	F	T

We also have the following:

A	$\neg A$
T	F
F	T

and

A	B	$A \vee B$	$A \leftrightarrow B$
T	T	T	T
T	F	T	F
F	T	T	F
F	F	F	T

Now we can check that all our propositional connectives can be built up from $\neg$ and $\to$:

$$A \wedge B = \neg(A \to \neg B),$$

$$A \vee B = \neg A \to B,$$

$$A \leftrightarrow B = (A \to B) \wedge (B \to A)$$

$$= \neg((A \to B) \to \neg(B \to A)).$$

Hence if we only introduce the connectives $\neg$ and $\wedge$ and ensure that they behave properly by suitable axioms, our logic will be suitably equipped with propositional connectives. We must, of course, require that these connectives preserve well-formedness.

2 Quantifiers. In setting up the theory of numbers, we must have at our disposal numerical variables. We thus introduce a sequence $x_1, x_2, x_3, \ldots$ of variables, each able to assume values that include the natural numbers.

Given a wff A, we wish to be able to assert that it holds for *all* possible values of one of its variables x_1, say, or that it holds for *at least one* value of a variable. We thus want:

Universal quantification: $(\forall x_i)A$: A is true for all values of x_i
Existential quantification: $(\exists x_i)A$: There exists at least one value of x_i for which A is true

We can express existential quantification in terms of universal quantification: to say that there exists an x_1 for which A is true is to deny that $\neg A$ is true for every x_1:

$$(\exists x_1)A = \neg(\forall x_1)\neg A.$$

If a variable in a wff is quantified by a universal or existential quantifier, we say that it is *bound*. Otherwise, we say that the variable is *free*. For example, x is bound in the (true) statement $(\forall x)[x + 2 > x]$, but is free in the predicate $[x + 2 > x] \& (\forall y)[y > 0]$. We say a wff is *closed* if no variable is free in it. Thus, in the intended interpretation, closed wff's represent sentences, whereas predicates are represented by wff's that are not closed. If we have a predicate, we may form a sentence from it by universal quantification over all its free variables. Thus $[x > y]$ yields the (false) sentence $(\forall x)(\forall y)[x > y]$. The closed wff $C(W)$ obtained from a given wff W is called the *closure of W*. Furthermore, if X is a wff and M is a variable free in X, we wish to be able to make *substitutions* and so denote by $S(X, M, N)$ the word that results on replacing the variable M by N at all of M's occurrences in X.

In general, predicate logic will need both constants and variables, as well as symbols to describe a variety of functions and predicates of different arities. However, we shall just consider a language $L(c_0, \ldots; f_0 \ldots; p_0, \ldots)$ with a set $\{c_0, \ldots\}$ of constant symbols, $\{x_0, \ldots\}$ of variables, $\{f_0, \ldots\}$ of unary function symbols, and $\{p_0, \ldots\}$ of binary function symbols.

3 Definition. We define the *terms* and *well-formed formulas* (*wffs*) of the language $L(c_0, \ldots; f_0, \ldots; p_0, \ldots)$ inductively as follows:
    We build up terms as follows:

(i) Each c_i is a term.
(ii) Each variable x_i is a term.
(iii) If t is a term, so too is $f_i(t)$ for each function symbol f_i.

We then build up wff's as follows:

(i) For each pair t_1, t_2 of terms, and each predicate symbol p_i, $p_i(t_1, t_2)$ is a wff.
(ii) If ϕ_1 and ϕ_2 are wff's, so too are $\neg\phi_1$ and $\phi_1 \to \phi_2$.
(iii) If ϕ is a formula and x_i is a variable, then $(\forall x_i)\phi$ is also a wff.

Note that quantification only applies to variables x_i. We do not allow quantification $(\forall f)$ over function symbols or $(\forall p)$ over predicate symbols. Because it precludes this higher-order quantification, we say that we are dealing with *first-order* logic.

4 Definition. By a *theory* Σ in first-order logic we simply mean a set of wff's of L. (We may think of these as axioms to be added to a standard set of logical axioms, and manipulated by a standard rule of inference, as given in the next definition.)

We must now introduce the notion of a proof in a theory Σ:

5 Definition. We say that the wff ϕ is *provable* with respect to Σ, and write $\Sigma \vdash \phi$, just in case:
   (a) ϕ is a wff in Σ:

$$\Sigma \vdash \phi \quad \text{for all } \phi \in \Sigma;$$

   (b) ϕ is obtained from one of the four axiom schemes below by substituting wff's for A, B and C:

   Axiom 1: $(A \to (B \to A))$
   Axiom 2: $((A \to (B \to C)) \to ((A \to B) \to (A \to C)))$
   Axiom 3: $((\neg B \to \neg A) \to (A \to B))$
   Axiom 4: $(S(A, x_i, c_j) \to (\exists x_i)A)$

where A is a wff, x_i is not bound in A, and c_j is a constant; or
   (c) ϕ is inferable from other provable statements by *modus ponens*: If $\Sigma \vdash \psi$ and $\Sigma \vdash (\psi \to \phi)$, then also $\Sigma \vdash \phi$.

Given a theory Σ and a wff ϕ we use $\Sigma + \phi$ to denote the theory $\Sigma \cup \{\phi\}$ obtained by adjoining ϕ to the set of wff's in Σ.
   Given a set $\Sigma = \{\phi_1, \ldots, \phi_n\}$ of wff's, we may note that, in terms of the wff's provable from it, it is equivalent to the single wff $\phi_1 \wedge \cdots \wedge \phi_n$.*
   This is because it follows from axioms 1 to 3 that

---

* Since our account here is not rigorous, we are being a little informal. $\phi_1 \wedge \phi_2 \wedge \phi_3$, for example, is shorthand for $((\phi_1 \wedge \phi_2) \wedge \phi_3)$. It is an exercise to verify that $\vdash ((\phi_1 \wedge \phi_2) \wedge \phi_3) \to (\phi_1 \wedge (\phi_2 \wedge \phi_3))$, and so on in the general case, $n \geq 3$, for each rearrangement of parentheses. We can thus use the shorthand $\phi_1 \wedge \cdots \wedge \phi_n$ with impunity.

$$\phi_1 \wedge \cdots \wedge \phi_n \vdash \phi_j \quad \text{for each } j,$$

so that if $\Sigma \vdash \phi$ for any ϕ, we can replace each invocation of $\Sigma \vdash \phi_j$ in the proof of ϕ by the steps of the proof that $\phi_1 \wedge \cdots \wedge \phi_n \vdash \phi_j$ to obtain a proof of $\phi_1 \wedge \cdots \wedge \phi_n \vdash \phi$. Conversely, it is easy to show that $\Sigma \vdash \phi_1 \wedge \cdots \wedge \phi_n$, so that any proof of $\phi_1 \wedge \cdots \wedge \phi_n \vdash \phi$ immediately yields a proof of $\Sigma \vdash \phi$.

Thus, in our general definition of a theory, we may think of Σ interchangeably as a set $\{\phi_1, \ldots, \phi_n\}$ of wff's, or as a single wff $\phi_1 \wedge \cdots \wedge \phi_n$. It should next be clear that

6 Observation. $\Sigma \vdash \phi$ *iff* $\vdash \Sigma \to \phi$ *where* $\vdash \psi$ *is shorthand for* $\varnothing \vdash \psi$, *with* $\varnothing$ *the empty set, i.e.,* ψ *is provable just from Axioms* 1 *through* 4.

Certainly, $\Sigma \vdash \Sigma$, and if $\vdash \Sigma \to \phi$, then $\Sigma \vdash \Sigma \to \phi$. Thus, by modus ponens, we deduce that $\Sigma \vdash \phi$.

Conversely, if $\Sigma \vdash \phi$ we simply note that *every* step of the proof, $\Sigma \vdash \psi_j$, can be replaced by $\vdash \Sigma \to \psi_j$. For example, suppose we have $\Sigma \vdash (\psi_i \to \psi_j)$ and $\Sigma \vdash \psi_i$ and have used modus ponens to deduce that $\Sigma \vdash \psi_j$. Instead, we now have $\vdash (\Sigma \to (\psi_i \to \psi_j))$ and $\vdash (\Sigma \to \psi_i)$. We use axiom 3 to obtain

$$\vdash ((\Sigma \to (\psi_i \to \psi_j)) \to ((\Sigma \to \psi_j) \to (\Sigma \to \psi_j))),$$

then one application of modus ponens yields

$$\vdash ((\Sigma \to \psi_j) \to (\Sigma \to \psi_j))$$

and then with another we obtain, as desired

$$\vdash (\Sigma \to \psi_j).$$

7 Definition. A wff W is called n-ary if the variables $x_1, \ldots x_n$ are free in W, and if no other variable is free in W. If W is an n-ary wff, and if $(k_1, \ldots, k_n)$ is an n-tuple of constants, we write $W(k_1, \ldots, k_n)$ to denote the wff obtained from W on replacing x_i by k_i at all occurrences of x_i in W, $1 \le i \le n$.

8 Definition. A logic Σ is consistent if for no wff ϕ do we have both $\Sigma \vdash \phi$ and $\Sigma \vdash \neg\phi$.

9 Lemma. *L is not consistent iff all wff's are theorems.*

PROOF. Suppose $\Sigma \vdash \phi$ and $\Sigma \vdash \neg\phi$. Let ψ be any wff. We wish to show $\Sigma \vdash \psi$.

By axiom 1, $\Sigma \vdash (\neg\phi \to (\neg\psi \to \neg\phi))$.
By modus ponens, $\Sigma \vdash (\neg\psi \to \neg\phi)$.
By axiom 2, $\Sigma \vdash ((\neg\psi \to \neg\phi) \to (\phi \to \psi))$.
By modus ponens $\Sigma \vdash (\phi \to \psi)$, and hence $\Sigma \vdash \psi$.

The converse is, by definition, trivial.   □

10 Lemma. *If* $\Sigma \not\vdash \neg\phi$, *then* $\Sigma + \phi$ *is consistent.*

PROOF. If $\Sigma + \phi$ were not consistent, then

$$\vdash (\Sigma \wedge \phi) \to \neg\phi.$$

In what follows we exploit (without proof) the well-known fact that any manipulation sanctioned by the truth tables can be proved from the axioms. We may thus transform the above assertion to

$$\vdash \neg(\Sigma \wedge \phi) \vee \neg\phi$$

i.e., $\vdash \neg\Sigma \vee \neg\phi \vee \neg\phi$
i.e., $\vdash \neg\Sigma \vee \neg\phi$
i.e., $\vdash (\Sigma \to \neg\phi)$
i.e., $\Sigma \vdash \neg\phi$.

The result follows by contradiction.   □

A famous set of axioms for the natural numbers, $\mathbf{N} = \{0, 1, 2, 3, \dots\}$ was given by the Italian mathematican Giuseppe Peano. The Peano axioms are as follows:

1. 0 is a natural number.
2. Every natural number x has a unique successor $s(x)$.
3. If x, y are natural numbers with $s(x) = s(y)$, then $x = y$.
4. For each natural number x, $s(x) \neq 0$.
5. If S is any subset such that (i) $0 \in S$ and (ii) if $x \in S$, then $s(x) \in S$, then $S = \mathbf{N}$.

With this, let us check that each of Peano's axioms can be expressed as a formula of the language of Definition 3:

1. $(\exists x) p_0(x, c_0)$.
2. $(\forall x)(\exists y)(p_0(y, f_0(x)))$.
3. $(\forall x)(\forall y)(p_0(f_0(x), f_0(y)) \to p_0(x, y))$.
4. $(\forall x) \neg p_0(f_0(x), c_0)$.
5. $(f(c_0) \wedge \forall x(f(x) \to f(s(x)))) \to (\forall x) f(x)$.

Here c_0 is to be interpreted as the integer 0, p_0 is to be interpreted as the predicate of numerical equality, f_0 is to be interpreted as the successor function, and f is a unary predicate variable. This motivates our general definition of a model (relational structure) and an interpretation:

11 Definition. We say $M = (A; b_0, \dots; g_0 \dots; p_0, \dots)$ is a *relational structure* or *model* for the language $L(c_0, \dots; f_0, \dots; p_0, \dots)$ with a set $\{c_0, \dots\}$ of constant symbols, $\{f_0, \dots,\}$ of function symbols, and $\{p_0, \dots\}$ of predicate symbols if

A is a set

each b_i is an element of *A*

each $g_i: A \rightarrow A$ is a (partial) function on *A*

each $q_i \subset A \times A$ is a binary relation on *A*.

The *M-interpretation* I of a term in L is defined by interpreting each constant c_i as the element b_i; and each f_i as the function $g_i: A \rightarrow A$. We thus have,

$$I(c_i) = b_i \text{ for each } i; \text{ while}$$

if t is a term with interpretation $I(t)$, then $f_i(t)$ is a term with interpretation $I(f_i(t)) = g_i(I(t))$.

We then say that M *satisfies* ϕ, or ϕ is *valid* in M, abbreviated $M \models \phi$ inductively for formulas ϕ if ϕ is "true" under the above interpretation. More formally, we have:

If t_1, t_2 are terms, $M \models p_i(t_1, t_2)$ just in case $I(t_1)q_i I(t_2)$.

If ϕ is a closed wff, then $M \models \neg\phi$ just in case it is not true that $M \models \phi$.

If ϕ_1 and ϕ_2 are closed wff's then $M \models (\phi_1 \wedge \phi_2)$ just in case both $M \models \phi_1$ and $M \models \phi_2$.

If ϕ is a wff and x is a variable such that $(\forall x)\phi$ is a closed wff, then $M \models (\forall x)\phi$ just in case $M \models \phi(a)$ for each choice of a in A.

12 Lindenbaum's Lemma. *If Σ is consistent, then Σ is a subset of a full set $\tilde{\Sigma}$, i.e., for each ϕ, either ϕ or $\neg\phi$ is in $\tilde{\Sigma}$.*

PROOF. Enumerate the wff's of L in some order as $\phi_0, \phi_1, \ldots$. Then define the sequence $\Sigma_0, \Sigma_1, \ldots, \Sigma_n, \ldots$ inductively by

$$\Sigma_0 = \Sigma,$$

$$\Sigma_{n+1} = \begin{cases} \Sigma_n & \text{if } \Sigma_n \vdash \neg\phi_n, \\ \Sigma_n + \phi_n & \text{if not.} \end{cases}$$

We check consistency by induction. Σ_0 is consistent by hypothesis on Σ. Suppose Σ_n is consistent. If $\Sigma_{n+1} = \Sigma_n$, then Σ_{n+1} is consistent. But, by Lemma 10, if $\Sigma_n \nvdash \neg\phi_n$, then $\Sigma_{n+1} = \Sigma_n + \phi_n$ is also consistent.

Now let $\tilde{\Sigma} = \bigcup_{n \geq 0} \Sigma_n$. This must also be consistent, since a proof of inconsistency must be finite and so lie within some Σ_n.

It is clear that $\tilde{\Sigma}$ is full. $\qquad\square$

Why doesn't Lindenbaum's Lemma contradict Gödel's Incompleteness Theorem? Well, for one thing, the above proof is not constructive, so there is no claim that $\tilde{\Sigma}$ is recursive or recursively enumerable, only that it exists. In our classic case of arithmetic, Σ can be any subset of the true statements about arithmetic, while $\tilde{\Sigma}$ could be the set of *all* true statements about arithmetic.

Note also that $\tilde{\Sigma}$ depends on the ordering $\phi_0, \phi_1, \ldots$. Suppose that Σ is incomplete, with neither $\Sigma \vdash \phi$ nor $\Sigma \vdash \neg\phi$. Then if $\phi_0 = \phi$, $\phi_1 = \neg\phi$, we would get $\phi \in \tilde{\Sigma}$, while if $\phi_0 = \neg\phi$ and $\phi_1 = \phi$, we get $\neg\phi \in \tilde{\Sigma}$.

13 The Gödel–Henkin Theorem. *If Σ is consistent, then Σ has a model, i.e., there exists M such that $M \models \Sigma$.*

PROOF. Take as our set A of constants the union of the set of all constants appearing in Σ with the disjoint sets $\{b_0, b_1, \ldots\}$ and $\{b_0', b_1', \ldots\}$.

Let $\{t_0, t_1, t_2, \ldots\}$ be an enumeration of all the terms with no free variables. Now enlarge Σ by adding to it all the wffs

$$((\exists x)(x = t_j) \to (b_j = t_j)), \qquad j = 0, 1, \ldots.$$

Fix a variable v_1, enumerate all the wff's containing v_1 as the only free variable as $\{\psi_0, \psi_1, \ldots\}$, and then add to Σ all the wff's

$$(\exists v_1)\psi_k(x_1) \to \psi_k(b_k'), \qquad k = 0, 1, \ldots.$$

Let then Σ^∞ be the set of all wff's obtained from Σ in this way. Σ^∞ is consistent since each b_j and b_k' is a new constant. We may thus apply Lindenbaum's Lemma to extend Σ^∞ to a full $\tilde{\Sigma}$, i.e., for each wff ϕ, either $\tilde{\Sigma} \vdash \phi$ or $\tilde{\Sigma} \vdash \neg\phi$.

We now define a model M that has as underlying set the set A of constants defined above, so that we interpret each constant directly as an element of A.

We then define the interpretation of the function symbol f_i to be the map $g_i \colon A \to A$ satisfying $g_i(c) = b_k'$ iff $(f_i(c) = v_1)$ was the kth wff in our enumeration above, for we then added $(f_i(c) = b_k')$ to our stock of assertions in Σ^∞.

We then define $q_j \subset A \times A$ to correspond to the binary predicate symbol p_j by setting

$$b g_j \hat{b} \text{ iff } \tilde{\Sigma} \vdash p_j(b, \hat{b}).$$

It follows from the construction that $M \models \phi$ if and only if $\tilde{\Sigma} \vdash \phi$. But since $\Sigma \subset \tilde{\Sigma}$, we thus have that $M \models \phi$ for every $\phi \in \Sigma$. Thus Σ has a model. In fact it is a *countable* model. $\square$

Having done all the hard work, it is now an easy matter to deduce

14 Gödel's Completeness Theorem. *If ϕ is satisfied by every model (we write this as $\models \phi$), then ϕ is provable, $\vdash \phi$.*

PROOF. Suppose that $\nvdash \phi$. Then by the Lemma, $\{\neg\phi\}$ is consistent and so, by Gödel–Henkin, has a model M. But then M cannot also satisfy ϕ, and so it is not true that $\models \phi$: not $\vdash \phi$ implies not $\models \phi$. Thus $\models \phi$ implies $\vdash \phi$. $\square$

We close this section by briefly returning to Gödel's *in*completeness theorem.

The first notion is that of *Gödelization*, that of treating any wff of L as if it were a number. For example, suppose we typed all our wff's on a keyboard with 50 keys, thus having access (with shift and no-shift) to exactly 100 characters. Then we could easily code each character as one from the list a_{00}, a_{01}, $a_{02} \ldots, a_{98}, a_{99}$. But then any string of characters can be given by a decimal number of even length. To avoid the problem of leading zeros, just prefix the string with 11, then read off each successive pair of digits ij to learn the corresponding character a_{ij}. (Actually, Gödel used a far more costly coding scheme, based on prime factorization, but it had pleasing numerical properties.)

Given this coding, it is pretty clear that we can write computer programs to check if a string of symbols is a single wff or a string of wff's. Another program can then check if a string of wff's is a valid proof or not. But Gödel's key observation was that all these programs (as we would now call them) could be seen as computing arithmetic functions of the kind that any self-respecting arithmetical logic must be able to describe. In particular, we could define the following numerical functions:

$$enc: N \to N \text{ with } enc(n) = \begin{cases} 1 & \text{if } n \text{ encodes a wff,} \\ 0 & \text{if not;} \end{cases}$$

$$h(n) = \begin{cases} 1 & \text{if } n \text{ encodes a list of wff's,} \\ 0 & \text{if not;} \end{cases}$$

$$h(m, n, j) = \begin{cases} 1 & \text{if } m \text{ encodes a list of wff's, and } n \text{ is the numerical} \\ & \text{encoding of the } j\text{th wff on the list,} \\ 0 & \text{if not.} \end{cases}$$

Given a finite list of axioms $A_1, \ldots, A_s$, we can define the function A which expresses the predicate "n encodes an axiom" by taking

$A(n)$
$$= \begin{cases} 1 & \text{if } n \text{ encodes a wff that is an instance of the axiom } A_i, \text{ for } 1 \le i \le s, \\ 0 & \text{if not.} \end{cases}$$

Again, given a rule of inference, $R(Y, X_1, \ldots, X_k)$, we can define a numerical function g_R by

$$g_R(m, n_1, \ldots, n_k) = \begin{cases} 1 & \text{if } m, n_1, \ldots, n_k \text{ encode wff's } Y, X_1, \ldots, X_k, \\ & \text{respectively, for which } R(Y, X_1, \ldots, X_k) \text{ holds,} \\ 0 & \text{if not.} \end{cases}$$

I do not say it is *easy* to write out the numerical encoding g_R of an R, only that it should be clear that it can be done.

Now let us introduce the predicate $Ded_R(m, n, j)$ to encode the assertion that n encodes the jth wff in the list encoded by m, and that this wff is deducible by the rule of inference R from wff's occurring *earlier* on the list:

$$Ded_R(m,n,j) = \begin{cases} h(m,n,j) \\ \wedge\ (\exists x_1)\cdots(\exists x_k)(\exists n_1)\cdots(\exists n_k) \\ ((1 \le x_1 < j) \wedge \cdots \wedge (1 \le x_k < j)) \\ \wedge\ (h(m,n_1,x_1) \wedge \cdots \wedge h(m,n_k,x_k)) \\ \wedge\ g_R(n,n_1,\ldots,n_k). \end{cases}$$

Let then $R_1,\ldots,R_r$ be our fixed set of rules of inference. We can then build up the function pr for which

$$pr(m) = \begin{cases} 1 & \text{if } m \text{ encodes a proof} \\ 0 & \text{if not} \end{cases}$$

very quickly as follows:

$$pr(m) = h(m) \wedge (\forall n)(\forall j)(h(m,n,j)$$
$$\rightarrow A(n) \vee Ded_{R_1}(m,n,j) \vee \cdots \vee Ded_{R_r}(m,n,j))).$$

Then to say that a wff is provable is encoded by

$$th(n) = (\exists m)(\exists j)(h(m,n,j) \wedge pr(m)).$$

i.e., n encodes the jth step in a valid proof encoded by m.

In Section 8.2 we said (somewhat less formally) that a recursive logic Σ was *adequate* if, for each recursively enumerable set U there corresponded a wff $U(x)$ with one free variable x such that for each numeral n

$$\Sigma \vdash U(n) \text{ if and only if } n \in U.$$

If Σ is *consistent*, then $\Sigma \vdash \neg U(n)$ entails that $\Sigma \nvdash U(n)$ and so $n \notin U$. Σ is *complete* if, further, for every U the converse also holds i.e., $\Sigma \vdash U(n)$ for each $n \in U$, while $\Sigma \vdash \neg U(n)$ for each $n \notin U$.

15 Observation. *An equivalent definition of adequacy of Σ is that for every recursive function $\phi: N^k \to N$, there is a predicate $\hat{\phi}(n_1,\ldots,n_k,n)$ which represents ϕ in Σ, i.e.,*

$$\Sigma \vdash \hat{\phi}(n_1,\ldots,n_k,n) \quad \text{if and only if } \phi(n_1,\ldots,n_k) = n.$$

The proof proceeds from the observation that, if we pick a coding of $(k+1)$-tuples of natural numbers as a single number, say $c(n_1,\ldots,n_k,n)$, then

$$\{c(n_1,\ldots,n_k,n) \mid \phi(n_1,\ldots,n_k) = n\}$$

is a recursively enumerable set.

We need one more change:

16 Definition. We say a logic is *ω-consistent* if not only is it consistent, but whenever $\Sigma \vdash (\exists x)W(x)$, then it is the case that there exists an actual numeral n for which $W(n)$ is true.

Thus our considerations to date make it clear that if Σ is adequate, then there is a predicate $Pf(x, y)$ such that

$\Sigma \vdash Pf(x, y)$ iff x is the Gödel number of a proof of the wff encoded by y.

In the same way, there is a predicate $Pf^+(x, y, z)$ such that

$$\Sigma \vdash Pf^+(x, y, z) \text{ iff}$$

(a) y encodes a wff with one free variable, and
(b) x is the Gödel number of a proof of the formula obtained by substituting the numeral for z for the free variable in that wff.

Now consider the wff

$$G = \neg(\exists x)Pf^+(x, y, y)$$

and let it have Gödel number g, with corresponding numeral $\bar{g}$.

Now consider the formula $\hat{G} = \neg(\exists x)Pf^+(x, \bar{g}, \bar{g})$. It asserts that

(a) $\bar{g}$ does encode a wff with one free variable, namely, $G = \neg(\exists x)Pf^+(x, y, y)$.
(b) There is *no* x which is the Gödel number of a proof of the formula obtained by substituting the numeral $\bar{g}$ for the free variable in G—but this is just $\hat{G}$ itself.

In other words, $\hat{G}$ "is" the assertion "$\hat{G}$ is not provable."
More formally, we have shown that

$$\Sigma \vdash \hat{G} \text{ iff } \hat{G} \text{ is not provable in } \Sigma.$$

Clearly, we cannot have $\Sigma \vdash \hat{G}$, for then we would have that $\hat{G}$, a contradiction.

Suppose, on the other hand, we have that $\Sigma \vdash \neg\hat{G}$, in other words that we can prove $(\exists x)Pf^+(x, \bar{g}, \bar{g})$.

Then by ω-*consistency*, there actually is a number m for which $Pf^+(m, \bar{g}, \bar{g})$, i.e., for which m encodes the proof of the wff encoded by g when $\bar{g}$ is substituted for its free variable. But then $\hat{G} = \neg(\exists x)Pf^+(x, y, y)$ *is* provable, $\Sigma \vdash \hat{G}$, contradicting our initial assertion that $\Sigma \vdash \neg\hat{G}$. In other words, neither $\Sigma \vdash \hat{G}$ nor $\Sigma \vdash \neg\hat{G}$. We have thus proved

Gödel's Incompleteness Theorem 2. *Every adequate ω-consistent arithmetical logic is incomplete.*

8.4 Speed-Up and Incompleteness

We start with a speed-up theorem, stated by Gödel (1936), and given in slightly modified form a proof by Mostowski (1957) in his exposition of Gödel's incompleteness theory. In Mostowski's monograph, the basic notions of addition, equality, and ordering of the natural numbers are formalized in first-order logic to yield a system Σ. It is shown that Σ is undecidable, and in fact an actual statement ψ is exhibited such that neither ψ nor its negation are provable within Σ if Σ is consistent. A new system Σ_1 is then formed, essentially by adjoining to Σ the axioms of second-order quantification, in which ψ is provable.

For any wff ϕ for which $\Sigma \vdash \phi$ (i.e., ϕ is provable in Σ), let $p(\phi)$ denote the length of the shortest proof of ϕ in Σ. Define $p_1(\phi)$ similarily for ϕ provable in Σ_1. Since Σ_1 is an extension of Σ, any Σ-proof of ϕ is an Σ_1-proof of ϕ, and so we immediately conclude that $p_1(\phi) \leq p(\phi)$ for every ϕ provable in Σ. However, more turns out to be true.

1 The Gödel Speed-Up Theorem. *For every recursive function F there exists a wff ϕ of Σ such that $\Sigma \vdash \phi$ and $\Sigma_1 \vdash \phi$ and such that the minimal Σ_0- and Σ_1-proofs of ϕ satisfy the inequality $p(\phi) \geq F(p_1(\phi))$.*

Mostowski's proof is based on a study of specific properties of the two systems (all the axioms of which are known), and uses results obtained in proving the completeness of Σ. Instead, let us give a more general approach (Arbib, 1966). We start by capturing the most abstract properties of a recursive logic for which we have some measure on the difficulty of proofs:

2 Definition. A PM-system (proof-measure system) on the alphabet X is a quadruple $L = (G, \phi, p, t)$, where

1. G is a recursive set of words on the alphabet X, the members of which are called well-formed formulas (wff).
2. A recursive enumeration (ϕ_1, p_1), (ϕ_2, p_2), (ϕ_3, p_3), ... Here, $\phi_n \in G$ is called the nth *proof*, and $p_n \in N$ is called the *measure* of the nth proof. (In this abstraction, we only list the result of the proof, and the length of the proof, not the intermediate steps of the proof.)

   We say ϕ is a *theorem* of L, $L \vdash \phi$, if and only if $\phi \in T_L = \{\phi_n | n = 1, 2, 3, \ldots\}$. By definition, $p(\phi) = \min\{p_n | \phi = \phi_n\}$, and this is defined only on T_L. $p(\phi)$ is the minimal proof measure of ϕ, and is called the *difficulty* of ϕ (in L).
3. t is an increasing total recursive function t such that

$$n > t(m) \Rightarrow p_n > m.$$

(This captures the idea that short proofs get listed earlier then much longer proofs.)

N.B.: (3) implies that if $m = p(\phi)$, then $\phi = \phi_n$ for some $n \leq t(m)$. In particular, $\min\{n | \phi = \phi_n\} \leq t(p(\phi))$ if $\phi \in T_L$.

Our intuitive idea of a speed-up is as follows. Take any function, as big as you please, say $r(x) = 2^{2^x}$. Then certainly if a theorem ϕ has difficulty a in one proof system L, and difficulty b in proof system L_1, we would concede that there is a real speed-up in going from a to b if $2^{2^b} \le a$. The next definition is even more dramatic. It says that no matter how big r may be, there is a ϕ^r whose proof is sped up by r:

3 Definition. Let L and L_1 be two PM-systems, with difficulties p and p_1, respectively. We say that L_1 is a *speed-up* of L if

(a) $T_L \subset T_{L_1}$.
(b) For every total recursive function r, L_1 is an r speed-up of L; i.e., there exists $\phi^r \in T_L$ such that $p(\phi^r) \ge r(p_1(\phi^r))$.

4 Lemma. *Let L be a PM-system, and let H be any infinite recursive subset of X^*. Then $L|H =_{def} (G \cap H, \psi, q, t)$ is a PM-system, where*

(a) $\psi_n = \phi_{\alpha(n)}$, where $\alpha(-1) = 0$, $\alpha(m+1) = \min\{n > \alpha(m)|\phi_n \in H\}$.
(b) $q_n = p_{\alpha(n)}$.

PROOF. We have only to verify (3) of the definition. But

$$n > t(m) \Rightarrow \alpha(n) > t(m) \Rightarrow p_{\alpha(n)} > m \Rightarrow q_n > m. \qquad \square$$

We now easily obtain a theorem which, roughly speaking, says that if a system L_1 can do some things arbitrarily quicker than L can, then L_1 can do some things that L cannot do at all.

5 Theorem. *Let L_1 be a speed-up of L. Then there is a $\theta \in G \cap G_1$ such that*

$$\theta \in T_{L_1} - T_L.$$

PROOF. Let $\hat{L}_1 = L_1|G = (G \cap G_1, \psi^1, q^1, t^1)$:

$$\hat{L} = L|G_1 = (G \cap G_1, \psi, q, t).$$

Clearly, $\hat{L}_1$ is a speed-up of $\hat{L}$.
   Let us assume, by way of contradiction, that $T_{L_1} \subset T_L$. Then we may define the total recursive function g by $g(n) = q_{s(n)}$, where

$$s(n) = \min\{m|\psi_n^1 = \psi_m\}.$$

Now set $\bar{g}(m) = \max\{g(n)|n \le t^1(m)\}$. Since $\hat{L}_1$ is a speed-up of $\hat{L}$, we may pick $\phi \in T_L$ such that

$$\bar{g}(g^1(\phi)) < q(\phi).$$

Let $m_0 = \min\{m|\phi = \psi_m\}$, $n_0 = \min\{n|\phi = \psi_n^1\}$, so that $q_{m_0} = g(n_0)$. But

$$q_{m_0} \ge q(\phi)$$
$$> \bar{g}(q^1((\phi)) \text{ by choice of } \phi$$
$$> g(n_0) \text{ by definition of } \bar{g}.$$

A contradiction! Thus there exists $\theta \in T_{\hat{L}_1} - T_{\hat{L}}$, whence $\theta \in T_{L_1} - T_L$. $\quad\square$

We now turn to the converse result. In 1964 I conjectured, having obtained the above results, that if we added a "missing statement" to an incomplete logic, then the resulting extension would be a speed-up. A proof was obtained by A. Ehrenfeucht and published with due acknowledgment in Arbib (1966); it was later republished by Ehrenfeucht and Mycielski (1971). To understand the setting for Ehrenfeucht's result, consider axiomatic systems with negation and implication satisfying the usual axioms of propositional logic.

6 Theorem (Ehrenfeucht). *Let* Σ *be a theory and* ϕ *a sentence, such that the* $\Sigma + \neg\phi$ *is undecidable. Then* $\Sigma + \phi$ *is a speed-up of* Σ.

PROOF. 1. By propositional logic

$$\Sigma + \neg\phi \vdash \theta \Leftrightarrow \Sigma \vdash \neg\phi \to \theta \Leftrightarrow \Sigma \vdash \neg\theta \to \phi.$$

Furthermore,

$$\Sigma + \phi \vdash \neg\theta \to \phi.$$

2. Assume, then, that $\Sigma + \phi$ is not a speed-up of Σ, so that there is a total recursive function g such that if $\Sigma \vdash \psi$, then $p_\Sigma(\psi) \le g(p_{\Sigma+\phi}(\psi))$.

Then we may effectively decide whether or not $\Sigma + \neg\phi \vdash \theta$ by searching the theorems of Σ with the proof measure $\le g(p_{\Sigma+\phi}(\neg\theta \to \phi))$ for $\neg\theta \to \phi$. This contradicts the undecidability of $\Sigma + \neg\phi$.   □

As a corollary to Ehrenfeucht's *proof* we obtain

7 Corollary. *Let* Σ *be a theory, and* ϕ *a sentence, such that* $\Sigma + \neg\phi$ *is undecidable. Then the difference set*

$$(\Sigma + \phi) - \Sigma$$

is not *recursively enumerable.*

PROOF. Bearing in mind (1) of the above proof, assume that $(\Sigma + \phi) - \Sigma$ is recursively enumerable. Then we may decide effectively whether or not $\Sigma + \neg\phi \vdash \theta$ by generating Σ and $(\Sigma + \phi) - \Sigma$ until we encounter $\neg\theta \to \phi$. Contradiction!   □

8.5 The Brain–Machine Controversy

Ernest Nagel and James R. Newman (1959) state that Gödel's theorem definitely limits the mathematical power of computers. Let me quote some of their concluding remarks:

> Gödel's conclusions bear on the question whether a calculating machine can be constructed that would match the human brain in mathematical intelligence. Today's calculating machines have a fixed set of directives built

into them; these directives correspond to the fixed rules of inference of formalized axiomatic procedure. The machines thus supply answers to problems by operating in a step-by-step manner, each step being controlled by the built-in directives. But, as Gödel showed in his incompleteness theorem, there are innumerable problems in elementary number theory that fall outside the scope of a fixed axiomatic method, and that such engines are incapable of answering, however intricate and ingenious their built-in mechanisms may be, and however rapid their operations. The human brain may, to be sure, have built-in limitations of its own ... [but Gödel's] theory does indicate that the structure and power of the human mind are far more complex and subtle than any non-living machine yet envisaged.

However, if our concern is with the general reach of machine intelligence, this argument seems to me to miss the point because (a) it accepts theorem-proving as the touchstone of human intelligence, and (b) it only applies to a mind-model that consists of a machine with all its knowledge carefully coded in logical form at some particular time and for which the only "mental operations" would consist of making strict deductions from the information encoded in it from the beginning. However, an adequate model of the mind would represent it as open to novel experience. One of the basic tasks in modeling mental activity is to understand our ability to learn from our mistakes. But if we make mistakes, we have certainly transgressed the limits of consistency. Since, as we saw in Chapters 4 and 5, machines have been built that incorporate learning algorithms, Gödel's theorem would seem to say nothing about their limitations.

It has been suggested that since Gödel's theorem shows that no device can enumerate all the truths of arithmetic without enumerating all the falsehoods as well, then we must be more intelligent than a machine. However, no human can discover all the truths of arithmetic and, of course, will be just as limited as a machine if he or she only discovers truths about arithmetic by sticking to a given set of axioms and a given set of rules of inference. We saw in Theorem 8.2.8 that there is a mechanical procedure for enlarging a given mechanical process for proving mathematical truths, and this already suggests that much of the force of the first argument against machine intelligence is vitiated by a deeper knowledge of the mathematical theory. It should also be noted that we have only considered machines that have a fixed structure, and we know that we may build machines which change their structure as a result of interaction with the environment. Thus, any truly intelligent machine would presumably have the ability to incorporate new "axioms" into its structure as a result of experience in numerical manipulation. Gödel's theorem limits a human as much as a machine; if a human closes his mind to new information, then there is much he cannot know. He can know much more if he continually modifies his ideas on the basis of comparing deductions from his own ideas with observations of the real world. So, any really adequate cybernetic theory of thought must contain a full understanding of growing inductive machines that continually change their internal structure as a result of interacting with the external environment.

Another striking observation about human thought is that we do not proceed from consistent axioms. Our internal model of the environment does contain a number of contradictions. When we do act logically, we try to proceed on the basis of this internal model in a way which is free from glaring contradictions. If we were to build a machine that interacted with the world in a way of any complexity, there would not be time to monitor all incoming information and compare it with all previously stored information in such a way as to ensure that there were no inconsistencies in the stored data whatsoever. Thus, for quite practical reasons we would expect occasional inconsistencies to arise, although presumably some process of editing would ensure that the inconsistencies did not become too prominent, if psychosis is to be avoided. In other words, if we do try to build a machine that is to interact with a complex environment in an ongoing way, it becomes impractical to make it a machine that operates completely consistently. At best, we can try to incorporate safeguards that will detect most mistakes as they arise and stop them from resulting in inadvisable action. Thus, we see another reason for rejecting the assertion that Gödel's theorem implies that machines cannot be intelligent, because *Gödel's theorem only places limitations upon consistent machines*, and tells us nothing about the design of machines, that are subject to inconsistency, albeit with various safeguards to prevent the overaccumulation of the inconsistencies in most situations.

I think that there is a really fascinating area of study awaiting the logician, namely, that of inconsistent systems in which we accept a proof if it does not take *too* long, and if it involves no *glaring* inconsistencies. I often have thought that judicial logic is a logic of this kind: The more unlikely the conclusion of the argument, the more expensive is the lawyer one must hire to convince a court of its truth.

Let us look at this in another way. Gödel's incompleteness theorem says that if you start with consistent axioms and apply the rules of inference, then the collection of theorems that can be deduced is incomplete in that not all true statements are provable. But there is some sense in which beings who live within the world are forced to be "complete," in that a decision as to which action is appropriate often cannot be postponed. If we are crossing the road and a car is bearing down upon us, we will not survive very long if we require a long period of careful theorem-proving before deciding whether or not it is true that the best action in this circumstance is to stay where we are or jump out of the way. Rather, as beings in the world, we must make many decisions with respect to certain real-time limitations. Thus, there is a sense in which time limitations force some form of completeness upon those statements that we must articulate in our actions.

We examine this further by considering a machine that, although still restricted unrealistically to analyzing well-formed formulas in a logical language, is now required, given such a formula, to print out "yes" or "no" within limited time to the question of whether the formula is a theorem, given certain axioms and rules of inference. Gödel's theorem implies that a machine that *only gives correct answers* must fail to answer certain questions—it will be incomplete. But, as the real world does to a human, we have imposed a time limit, within which the machine *must* deliver an answer "yes" or "no." In the human world, not to act is itself a decision. By forcing the machine to always reach a decision, we have in effect changed the rules of inference, e.g., by adding to the rules a running confidence level evaluator, and having the machine choose that answer "yes" or "no" which is more highly evaluated when the time is up, unless a complete proof has already yielded the answer. The new "logic" embodied in this machine is now complete—every question must receive an answer—so Gödel's theorem tells us that the logic must be inconsistent. In other words, time pressures guarantee that mistakes will be made. Hardly a result which separates man from machine!

Next let us modify our machine somewhat further, in a way that will give us insight into learning machines. When a mathematician has proved a theorem, he or she thenceforth treats it as if it were an axiom, i.e., as a truth from which other truths may be derived. Imagine that our machine is similarly equipped to add to its "data base" those statements it has evaluated to be true. Imagine, also, that its evaluations may result in effects upon the external world, such as the execution of an action whose appropriateness is asserted by the statement just "verified." In some case, no adverse consequences will follow, and so the assertion of the correctness of the action will appear to be confirmed. But if there are adverse consequences, the statement should be removed from the data base. But is this enough? Is it only the time constraint that led to the incorrect evaluation, or the consequence of false statements having entered the data base earlier? If the latter, and if many statements were invoked in the "proof" of the assertion just rejected, how are we to "assign blame" to weed the data base?

Humans do not live by rigorous arguments, but rather make decisions that seem plausible, living within a net of "elastic" entailments and continual interaction with their environment. Particular situations may force us to question specific entailments and change accordingly, but there is no guarantee that this will be for the better. We have seen that our decisions will, almost inevitably, by inconsistent taken as a whole. Certain relatively coherent and well-rehearsed domains will, of course, be "almost" free from contradictions. Given more time, we can seek more data, consider more alternatives, and thus extend the network of consistency (where we now use consistency in the sense of avoiding situations in which we seem to have compelling grounds both for and against a given course of action.) Mathematics is the limiting case in which time restrictions are removed. But even here, in the actual work of the mathematician, mistakes can be made, with subtleties omitted from a proof or intuitions overriding a fatal flaw in a logical argument.

Our everyday beliefs do not differ from mathematics only in the time constraints that attend their formation or application; nor in the fuzziness (limitation of consensus as to applicability) that attends them. Perhaps even more important is that a belief is not possessed of a simple binary indicator, "true" or "false." Rather [and this is reflected in the notion of "activation level" in schema theory (Arbib, Conklin and Hill 1987)], a belief may be more or less strongly held, and this "degree of belief" will vary with the circumstance. "Thou shalt not kill" holds strongly in the form "Thou shalt not kill humans" for most of us unless we are in a situation of war or self-defense, but does not prevent the carnivorous from ignoring the call of vegetarianism. There have been attempts to develop logics of belief in which each assertion is assigned a "degree of rational belief," the odds an omniscient gambler would take against an ingenious opponent with exactly the same information. However, just as Gödel's theorem blocks us from correctly evaluating the truth of all statements in a formal system, so it is that we, in finite time, cannot come up with exact degrees of rational belief. Still, theories of learning and of evolution can address ways that may (but cannot always) improve the odds of success or survival.

What about formal versus informal for the distinction between mechanical or human reasoning? I think formal depends on level of discourse: Human thought is informal at the level of English, formal at the biochemical level. Thus, a machine recognizing a letter A by weightings of an area of threshold elements (Chapter 4) cannot tell you what it is doing; it just "feels" the letter is an A. In the same way, we cannot formalize many of our arguments, and yet some superhuman intelligence might, in principle, show how our conclusion is a "logical consequence" of the state of our brain, and its interaction with the environment. We have such complex brains, and our receptors can receive so much information, that we cannot reduce all our mental transactions, influenced by emotional (brain-mediated and hormone-mediated) reactions, to the linguistic level. But presumably, each well-defined aspect of human mentation can be subsumed in a formalizable psychological–neurophysiological theory and, then, in principle, a machine could exhibit that activity. Perhaps Gödel's theorem says we can never build a machine that is exactly like a human being; reflection says it probably only tells us that if by chance we should build one, we might have no effective procedure for telling whether or not it really meets the criteria; and common sense says we would never want to build such a machine, since we build machines to help us, not to be us.

At this stage, we should emphasize a distinction we must make between thinking of computers in a purely symbol-manipulation sense, and thinking of them in the robotic sense. We say that a machine is a *general-purpose computer* if, like a universal Turing machine (Chapter 6), it can, given a suitable program and enough space and time, compute any function, as long as the data are provided in *suitably encoded form*, and the answer will be accepted in some coded form. This is quite different from the problem of the robot which cannot demand that the data be specifically encoded, but must receive sensory information from the energy in the world around it, and must respond by actually moving its effectors in and about the world. Thus, even if the central nervous system of a human is, hypothetically at least, a universal computer, it cannot function appropriately until it is coupled to appropriate systems for receiving light and sound energy from the world, and for producing speech or motor activity as required. Much of a child's development consists precisely of this matching of receptors and effectors to the central computation system. Without such matching no appropriate central programming can occur, and so the universality of the central computing network can only remain potential until such time as effectors and receptors are properly coupled.

Furthermore, even if a computer is general purpose, in that any computation can be carried out by an appropriate program in terms of its basic instructions, nonetheless, by adding new instructions we may increase the practical range of the computer, by making certain programs much more efficient. Of relevance here is the Gödel speed-up theorem (Section 8.4). It states that if we add an undecidable axiom to an incomplete logic, not only are there truths that become theorems for the first time, but also theorems that were already provable in the old system may have shorter proofs in the new system; in fact, given any criterion of what you mean by speeding up a proof, there will exist at least one, and in fact infinitely many, theorems whose proofs are sped up by this criterion. This accords well with the intuition of those who have studied measure theory, and who know how much easier proofs are when we make use of the axiom of choice, rather than trying to give a constructive proof which can be formalized within a system that does not include it. If we make the highly artificial assumption that the contents of the memory of an organism corresponds to the axioms, and theorems so far, of a formal system—with the ability to add new axioms given by induction—then this theorem reminds us that the virtue of adding new data to memory is not simply that that particular piece of information now becomes available to the organism, but also that other information may be computed by the organism much faster than it could otherwise, even though that information was not denied to the organism before. It suggests to me that an appropriate measure of information content for a statement is to be found in a measure of the reduction it effects in the computation required by the system when operating in a certain limited environment. In any case, we see that in looking at the design of memory structures, we must consider not only the speed of storing and retrieval of particular items, but we must also study ways in which we can choose what is to be stored in such a way as to most speed up the necessary computations of the system. In line with this, we shall want a precise understanding of the cumulative effects of small initial changes: one change favors certain other changes by making them relatively easy, and the next change in turn shifts the balance of probable computational structure, until a huge spectrum of styles prove to be consistent with an initial genetic structure, and we are selected between by virtually insignificant changes.

Thus, although the Gödel speed-up theorem is far removed from a realistic study of the brain, it indicates once again how a clear formal understanding of an area within automata theory can spur us toward asking much more incisive questions about the fundamental nature of information processing in the nervous system than might otherwise be possible.

In all this, we see our study of *Brains, Machines, and Mathematics* not as reducing humans to the level of current machines but rather as open-ended, expanding in response to critiques of the limitations of our current understanding of brain and machine.

References for Chapter 8

For a fuller and very readable discussion of the foundations of mathematics (Section 8.1), see Nagel and Newman (1959). For a thorough-going technical treatment, see Beth (1959).

In preparing Section 8.3, I found Crossley et al. (1972) very helpful, and recommend it to the reader who wants to learn more about mathematical logic from a short book (77 pages) that combines a semiformal treatment with a large helping of intuition. Among the many full textbook treatments is Ebbinghaus, Flum, and Thomas (1984).

Section 8.5 greatly expands the treatment of Arbib (1964) by incorporating material from the discussion of Gödel's incompleteness theorem in Arbib (1969), and Arbib and Hesse (1986). The reader may turn to either of these books for a far fuller exploration of the philosophical questions touched upon here.

Arbib, M.A., 1964, *Brains, Machines, and Mathematics*, McGraw-Hill.

Arbib, M.A., 1966, Speed-up theorems and incompleteness theorems, in *Automata Theory* (E.R. Caianiello, Ed.), Academic Press, pp. 6–24.

Arbib, M.A., 1969, Automata theory as an abstract boundary condition for the study of information processing in the nervous system, in *Information Processing in the Nervous System* (K.N. Leibovic, Ed.), Springer-Verlag, pp. 3–19.

Arbib, M.A., 1985, *In Search of the Person: Philosophical Explorations in Cognitive Science*, University of Massachusetts Press.

Arbib, M.A., Conklin, E.J., and Hill, J.C., 1987, *From Schema Theory to Language*, Oxford University Press.

Arbib, M.A. and Hesse, M.B., 1986, *The Construction of Reality*, Cambridge University Press.

Beth, E.W., 1959, *The Foundations of Mathematics*, North Holland Publishing Company.

Blum, M., 1967, A machine-independent theory of the complexity of recursive functions, *J.A.C.M.* **14**: 322–36.

Crossley, J.N., Ash, C.J., Brickhill, C.J., Stillwell, J.C., and Williams, N.H., 1972, *What is Mathematical Logic?*, Oxford University Press.

Davis, M., 1958, *Computability and Unsolvability*, McGraw-Hill.

Ebbinghaus, H.-D., Flum, J., and Thomas, W., 1984, *Mathematical Logic*, Springer-Verlag.

Ehrenfeucht, A., and Mycielski, J., 1971, Abbreviating proofs by adding new axioms, *Bull. Amer. Math Soc.* **77**: 366–367.

Gödel, K., 1931, Über formal unentscheidbare Sätze der *Principia Mathematica* und verwandter Systeme, I, *Monats. Math. Phys.* **38**: 173–198. [This is now available in English translation; see Kurt Gödel, *On Formally Undecidable Propositions of Principia Mathematica and Related Systems*, (translated by B. Meltzer, with an introduction by R.B. Braithwaite), Basic Books, Inc., Publishers, NY. Part II has not appeared.]

Gödel, K., 1936, Über die Länge der Beweise, *Ergeb. eines math. Kolloquiums* **7**: 23–24.

Mostowski, A., 1957, *Sentences Undecidable in Formalized Arithmetic*, North Holland.

Myhill, J., 1964, The abstract theory of self-reproduction, in *Views on General System Theory* (M.D. Mesarovic, Ed.), John Wiley & Sons, pp. 106–118.

Nagel, E. and Newman R., 1959, *Gödel's Proof*, New York University Press.

APPENDIX
Basic Notions of Set Theory

We think of a *set* as merely a collection of objects (called its *elements*), be they points, numbers, or the states of a finite automaton.

If X has the elements $x_1, x_2, x_3, \ldots$, we sometimes write $X = \{x_1, x_2, x_3, \ldots\}$.

By an extension of language, we include the notion of the *empty set*, i.e., a set with no elements. We denote it by $\emptyset$, so that $\emptyset = \{\}$.

We use the notation

$$x \in X$$

to denote that x belongs to the set X. For example, if X is the set of even numbers, then

$$2 \in X, \qquad -340 \in X, \qquad 0 \in X,$$

but 1 and 7, for example, do not belong to X, and so we write

$$1 \notin X, \qquad 7 \notin X.$$

We use the notation $\{x|y\}$ for "the set of all x for which the statement y is true." Thus:

$$X = \{x | x \in X\}.$$

The set of even numbers $= \{x_| = 2n \text{ for some integer } n\}$.

If X and Y are two sets, we write $X \subset Y$ to indicate that X is a *subset* of Y, i.e., that every element of X is also an element of Y.

For example, for any set of X, we have $\emptyset \subset X$. If X and Y are two sets, then

$$X \cap Y = \text{the } \textit{intersection} \text{ of } X \text{ and } Y$$

$$= \{x | x \in X \text{ and } x \in Y\}$$

$$= \text{the set of elements common to } X \text{ and } Y.$$

$$X \cup Y = \text{the } union \text{ of } X \text{ and } Y$$

$$= \{x | x \in X \text{ or } x \in Y\}$$

$$= \text{the set of elements belonging to } X \text{ or } Y \text{ or both.}$$

If the X_a are a collection of sets, we let

$$\bigcup \{X_a | P(a)\} = \{x | \text{there is an } a \text{ for which } P(a) \text{ is true and } x \notin X_a\}$$

$$= \text{the } union \text{ of those } X_a \text{ for which } P(a) \text{ is true.}$$

If X and Y are two sets, then the *difference* $X - Y = \{x | x \in X \text{ and } x \notin Y\}$.
Let $\mathbf{N} = \{0, 1, 2, 3 \dots\}$ be the set of natural numbers. If $X \subset \mathbf{N}$, then we
define the *complement* of X as

$$\bar{X} = \mathbf{N} - X$$

$$= \{x | x \text{ is a natural number not in } X\}.$$

We use the notation $X_1 \times X_2$ to denote set

$$\{(x_1, x_2) | x_1 \in X_1, x_2 \in X_2\},$$

that is, the set of all ordered pairs (x_1, x_2) of which the first element comes
from X_1 and the second from X_2. $X_1 \times X_2$ is called the *Cartesian product* of X_1
and X_2. For example, let R denote the set of all real numbers. Then when we
use Cartesian coordinates to specify the positions of points in the plane, we are
simply identifying the plane with the Cartesian product $R \times R$.

The next crucial concept is that of a *function f* which assigns an element in
one set Y to every element in another set X. We may say f maps the *domain X*
into the *codomain Y*, and denote this by

$$f: X \to Y.$$

We often write $f(x)$ for the element in Y to which f sends the element x in
X, and denote this by

$$x \mapsto f(x).$$

Thus, the arrow $\to$ tells us the sets with which f operates, while the barred
arrow $\mapsto$ gives the action of f on individual elements.

For example, the function of multiplication by 10 assigns a nonnegative
integer to every element of the set $\mathbf{N} = \{0, 1, 2, 3, \dots\}$ of nonnegative integers,
and so if we use $10(\cdot)$ to denote this function, with the $\cdot$ telling us where the
argument is to be inserted, we may write

$$10(\cdot): \mathbf{N} \to \mathbf{N}: x \mapsto 10x$$

to describe completely *where* and *how* this function functions. Often to describe the function (in the colloquial sense) of a biological system, we shall
need several functions in the mathematical sense.

We call a function f *onto* (or *surjective*) if f maps X onto Y in the sense that
for each y in its codomain Y, there is an x in its domain X for which $f(x) = y$.

We call a function *one-to-one* (or *injective*) if any one point in Y corresponds to at most one point in X, in that for *any* two distinct points $x \neq x'$ of X we have $f(x) \neq f(x')$.

Two functions f and g are said to be *equal* only when the following three conditions are all satisfied: (a) their domains are equal, (b) their codomains are equal, and (c) for each x in the domain, $f(x) = g(x)$.

The *composite* of f with g is defined only when the codomain of f is the domain of g, and is the function $x \mapsto g(f(x))$ on the domain of f to the codomain of g; the composite will be written $g \cdot f$ of gf, whichever is most convenient.

If X is a subset of A, then the characteristic function of X, $\chi_X \colon A \to \{0, 1\}$ is defined by

$$\chi_X(x) = \begin{cases} 0 & \text{if } x \notin X, \\ 1 & \text{if } x \in X. \end{cases}$$

Author Index

Subject Index